I0630078

Advance Praise for *Relentless Innovation*

Innovation isn't a "once and done" activity—you must develop a sustainable culture of innovation supported by a consistent process, tools, and training. For us, it's finding that balance between efficiency and innovation that makes innovation successful.

—Valerie Felice Cameo, Innovation U.S. Bank

The barriers to innovation are so woven into the fabric of "business as usual" that they're almost invisible. Jeffrey illuminates them and offers practical advice to drive sustainable innovation.

—Chuck Frey, InnovationTools

This is a book likely to burst much of the bubble that presently surrounds the innovation agenda in large organizations. While this may be uncomfortable for innovation professionals, the insights herein will go far to help leaders create innovation efforts which actually work. A recommended text for any executive with innovation ambitions.

—James Gardner, MD, International Spigit

Relentless Innovation tackles two really critical but surprisingly fundamental issues—one: innovation is not business as usual, and two: middle managers are not equipped today to make it be the case. Jeffrey Phillips opens up the black box of innovation and offers the blueprint that strikes the balance for the new organization in achieving a "dual capability" of achieving both efficiency and innovation. It all integrates into a consistent set of capabilities and he offers how innovation can finally become a core everyday discipline sustained by the middle manager, the new leader of relentless innovation.

—Paul Hobcraft, Agility Innovation & Paul4innovating.com

Jeffrey Phillips is one of the great mythbusters and truth tellers in the innovation space today. A respected voice with insights to share, Jeffrey does a great job in *Relentless Innovation* of examining many of the reasons that innovation does or does not take place within firms, and how corporate innovation efforts can get "stuck in the middle." If your organization is stuck and seeking to become a relentless innovator, this would be a great read.

—Braden Kelley, cofounder of InnovationExcellence.com and author of *Stoking Your Innovation Bonfire*

We are entering a new era in which innovation is managed. In this book, Jeffrey Phillips has captured the challenges facing organizations. As he points out: "No other important function in your business is ad hoc." With examples from relentless innovators, he identifies the two primary barriers to innovation and describes how to advance through and beyond them.

—Dr. John Lewis, Holosoft Corporation

Jeffrey Phillips nails two crucial factors for innovation success: (1) a rebalance of the operating model between efficiency and innovation, and (2) a refocus of middle managers toward innovation. *Relentless Innovation* provides excellent insights on how to nurture these preconditions in order to manage ambidexterity and stimulate innovation flow—essential reading for every innovator.

—Ralph-Christian Ohr, Ph.D., Product/Innovation Manager, Switzerland

Innovation is not a one-time thing. Creating a sustainable innovation culture has a lot of moving parts. Phillips does a great job of nailing down two of the most critical components.

—Dennis L. Potter, VP Innovation RJRT

Jeffrey Phillips has nailed the most perplexing phenomena that surround the management of innovation in complex corporate and institutional environments. Jeffrey's myths and truths couldn't be more insightful and useful for people who are committed to getting new products, services, and solutions to market—despite the organizational issues they will no doubt encounter along the way.

—Jeneanne Rae, President, Motiv Strategies

Even Steve Jobs wouldn't have been able to turn around Dell on his own. But armed with *Relentless Innovation*, you might have the tools to make innovation repeatable by leveraging your middle management. If you want to create a culture of innovation, you need this book!

—Stephen M. Shapiro, author, *Best Practices Are Stupid* and *Personality Poker*

Jeffrey Phillips's new book starts the myth-busting from page one onward and then constructs a new view of innovation as "business as usual" that should act as a blueprint for any executive team looking to bring the right combination of order and inspiration to its innovation efforts.

—Haydn Shaughnessy

How do successful innovators sustain innovation over a long period of time? This is the question that Phillips asks and answers in this valuable book. He demonstrates what it takes to make innovation "business as usual" in your organization. The book is packed with examples, tips, and secrets for innovation success.

—Paul Sloane, author of *The Innovative Leader*

RELENTLESS

RELENTLESS

RELENTLESS

RELENTLESS

RELENTLESS INNOVATION

WHAT WORKS, WHAT DOESN'T—AND WHAT THAT MEANS FOR YOUR BUSINESS

JEFFREY PHILLIPS

New York Chicago San Francisco Lisbon London Madrid Mexico City
Milan New Delhi San Juan Seoul Singapore Sydney Toronto

The McGraw·Hill Companies

1 2 3 4 5 6 7 8 9 0 DOC/DOC 1 6 5 4 3 2 1

ISBN 978-0-07-178680-5
MHID 0-07-178680-5

e-ISBN 978-0-07-178681-2
e-MHID 0-07-178681-3

This publication is designed to provide accurate and authoritative information in regard to the subject matter covered. It is sold with the understanding that neither the author nor the publisher is engaged in rendering legal, accounting, securities trading, or other professional services. If legal advice or other expert assistance is required, the services of a competent professional person should be sought.

—*From a Declaration of Principles Jointly Adopted by a Committee of the American Bar Association and a Committee of Publishers and Associations*

McGraw-Hill books are available at special quantity discounts to use as premiums and sales promotions or for use in corporate training programs. To contact a representative, please e-mail us at bulksales@mcgraw-hill.com.

This book is printed on acid-free paper.

To the best chauffeur I know, who also happens to be my spouse and my best friend. Catherine, thank you for everything you do.

Contents

Acknowledgments

When it comes to writing a book, there are many sources of inspiration and support. I'd like to recognize a few of mine.

First, there are the people who started out as clients and became good friends and colleagues. I think I've learned at least as much from them as they have learned from working with OVO®. A short list of those people includes Art Beckman, Oliver Burns, Valerie Cameo, Scott Edwards, Eric Fennel, Matt Gymer, Jeff Kroll, Todd Moning, Chad Pomeroy, Denny Potter, Lisa Shumway, and Dominic Venturo. Oliver introduced me to the concept of BAU, or business as usual. Working with these and other clients, I realized that BAU and middle managers are the biggest roadblocks to innovation, and they could also accelerate innovation if properly aligned and motivated. These individuals helped me to shape my thinking and helped to validate the ideas in this book.

The second source is the innovation community. Innovation is a relatively diverse field, yet it is still small enough that one can become acquainted with a number of people who work in this space. Authors like Tim Hurson, Roger von Oech, Stefan

Lindegaard, Steven Shapiro, Peter Schwartz, Gary Hamel, and Henry Chesbrough have created a valuable body of knowledge that many of us draw from. Innovation practitioners like Gene Slowinski, Paul Sloane, Paul Hobcraft, Tim Ogilvie, Jeneanne Rae, Chuck Frey, Braden Kelley, Rene Hopkins, Kathie Thomas, and Tim Kastelle are good friends who contribute significantly to the knowledge base about innovation. I've been fortunate enough to be associated with the Product Development Management Association (PDMA) and the Association of Managers of Innovation (AMI), two groups with a significant focus on innovation that have helped me extend my network and meet people who are driving the science of innovation forward.

The third source is my colleagues and partners at OVO and NetCentrics. Bob, Bob, and Dean were convinced that innovation, in the private sector as well as in the public sector, was a key differentiating factor. NetCentrics has demonstrated the ability to deliver innovation in federal contracts, due to the investments it made to help OVO become an innovation leader. Everyone at NetCentrics and OVO has supported my efforts to write this book. Thanks to everyone for your support and encouragement.

From these sources and others I gained insights that I hope to share with you. But getting those insights into a readable fashion required at least two other people I'd like to recognize: Niki Papadopoulos and Zach Gajewski, both at McGraw-Hill. Niki helped frame the theme of the book, and Zach trimmed, shaped, and directed my written wanderings into a much better book than it was at the start. Thanks to Niki for her vision and guidance and to Zach for making the book readable.

Most important, I'd like to recognize my family. Catherine, my wife, has shouldered a lot of the work at home raising three terrific kids while I've been living out of a suitcase working with clients. She has been a constant supporter of my work and my goal to write this book. My kids, Helen, Rachel, and Jacob, are partially responsible for this book. They are voracious readers, and for years I have encouraged them to write their own stories. They've encouraged me to write this one.

Introduction

What's unusual about business today is how usual it is. Most successful businesses "hum" with efficiency and effectiveness. Over the past two decades, business models have been finely tuned to crank out products that meet customer needs and consistently achieve quarterly forecasts. Through these highly predictive, carefully defined approaches, corporations imagine they've eliminated variance, errors, and surprises. Competitive forces, global markets, and consumer demands, however, are changing how firms compete by introducing far more variance. Customers and markets demand new products and services. To fulfill those demands, firms turn to innovation, which initiates a shift from known, consistent processes to unknown or unusual ones.

Since innovation is unfamiliar to so many businesses, it is often unsuccessful. Successful firms receive a significant amount of attention in the press when their products or services are disruptive or compelling and their methods differentiate them from others. The unusual nature of innovation and the lack of experience and knowledge about innovation methods, tools, and techniques, lead to a significant demand for information about innovation and innovative companies. We want to know

- First, how successful firms innovate
- Second, the secrets of their success
- Third, how we defend against innovators in our industry

Innovation happens in every sector of the economy, but its success is most consistently recognized in private enterprises. Increasingly, though, interest in innovation is moving beyond the private sector. Federal, state, and local governments are beginning to recognize the need to use innovation to produce more services with the same resources. As both the private and public sectors seek information and assistance to improve innovation capabilities, the demand for information about innovation will continue to grow.

Why Another Book on Innovation?

Given the increased interest in innovation tools and techniques it is no surprise that hundreds of books have been written about innovation. While many books have been written, the focus of those books has been fairly limited. To date, the vast majority of innovation books have been based on one of three formulas:

- How (famous company) innovates and what you can learn from that success
- Why your firm should implement (innovation technique) for growth and profits

- What you can learn from (innovation leader's) successes

If you've read one book based on each of these formulas, you've got a good introduction to innovation, and you may have learned from the experiences of leading companies if their situations and challenges are the same as yours. However, while you can learn a lot about how, say, Apple innovates, the techniques and methods Apple uses will not necessarily translate to your business or your challenges. What was successful for Apple may or may not be applicable for your firm. The market conditions, situations, and circumstances also may not reflect your firm's reality. Further, business models or solutions that Apple finds acceptable might not be acceptable in your business culture.

Your firm can implement innovation techniques documented by leading innovation experts if those techniques solve problems or challenges *that you face*. The range of innovation tools, techniques, and methods is vast, though, so not every tool or technique is applicable in every situation or industry. For example, firms that succeed using TRIZ may never need or use "open innovation." While these techniques can provide your firm with excellent results, you must understand the implications of the tools and be part of an organization that can implement the concepts and sustain their use over time.

Over the last decade, the public has been led to believe that powerful, insightful leaders make a significant difference in their firm's operations and profitability. But as even visionary leaders, like Jack Welch of GE, have demonstrated, leadership may not sustain innovation over time. A book providing information on how one executive implemented an innovation strategy in his or her business may not be applicable to your situation and needs.

Also, success in leadership in one firm often doesn't translate to success in another firm.

Further, few of these adulatory books examine in any depth the struggles or failures that were part of the innovation effort or the lessons learned from those failures. Most of these books focus only on the *successful* implementation of innovation, which represents a fraction of the collected knowledge about the subject.

What's Different?

Unlike the existing literature on innovation, the purpose of this book is to identify two simple yet powerful barriers that cause every firm to resist innovation. Those two barriers, properly motivated and aligned, can become the engine that drives innovation. I'll use some case studies to demonstrate important points, but I'm not going to focus on a specific company's inapplicable actions. I don't advocate specific tools in the text, like Systematic Inventive Thinking or "needs-based innovation" or trend spotting, since each of these is valid in some circumstances and not valid in others. And I'm not going to tell you to think like Steve Jobs, because, frankly, no one but Jobs can think that way.

I believe that the fundamental business models in most businesses have become unbalanced, sacrificing innovation for the sake of efficiency and effectiveness. Business models are so focused on efficiency, cost cutting, and short-term outcome, that it makes innovation almost impossible to accomplish once, much less over time. In this book, I explore the factors that sustain the imbalance in the model and I make recommendations to help your firm regain a balance between efficiency and innovation. Until that balance is achieved, your firm won't be a successful innovator, regardless of the tools you deploy or the experts you follow.

My Qualifications

For almost 10 years I have led the OVO Innovation consulting team, working with Fortune 500 and midsized firms in North America, Europe, and Asia. OVO Innovation was founded to help clients build and develop their internal innovation capabilities, with the goal of implementing innovation as a consistent business discipline. To that end we've worked with a number of companies to help define and train innovation teams, design and deploy innovation processes, and lead innovation projects. I wrote and published a book entitled *Make Us More Innovative*, targeting the individuals who are called on to lead innovation efforts. That book defined OVO's innovation approach and methodology, and it has been used successfully by our clients. I've been voted by innovation practitioners as one of the leading voices in the innovation community for my articles on the topic, especially those from my blog (Innovate on Purpose). I'm regularly asked to speak to corporate boards and conferences on innovation topics.

I've led innovation projects in leading firms in high technology, banking and finance, insurance, pharmaceuticals, and consumer goods, as well as government agencies. I've seen some substantial innovation successes—new compelling ideas brought to market, patents filed on new discoveries from our innovation processes—as well as some significant innovation failures.

I've led innovation workshops and training programs throughout the world and used techniques and tools that range from the most basic creativity skills to robust idea management software applications and, most recently, immersive experiences in 3-D virtual worlds with Second Life® as an ideation and rapid prototyping platform.

Through my consulting and training work, and my opportunities to interact with potential innovators, it became evident to me that many organizations want to innovate, but regardless of the tools, techniques, training, and consulting advice they received, they often failed to achieve this goal. The reasons for these failures, I decided, were structural and could not be addressed with the innovation tools and techniques we innovation consultants bring to bear. Until the structural issues are remedied, most organizations will experience, at best, occasional innovation, and at worst, will resist innovation.

Why Innovate Now?

Innovation is unusual in most businesses and therefore it takes low priority. The majority of companies don't realize how detrimental this lack of innovation is to their ability to grow and compete on a global basis. Innovation must make a transition from an "unusual" infrequent activity to a more common, consistent capability, regularly applied, to advance a company's visibility, culture, and profits—a concept that I call *relentless innovation.*

Relentless innovation is necessary for business success in the future. Firms can no longer innovate occasionally or halfheartedly. Today, we see three categories of firms: relentless innovators, occasional innovators, and everyone else. Relentless innovators are constantly generating new products and services. Several of these firms will play a prominent role in this book as I describe the factors that make them successful. Other firms innovate only occasionally, under severe market pressure, while some firms have developed strategies that led them to believe that they can survive by quickly following the innovators. Due to the dynamic forces of the global market, increasing consumer demand, and lower trade barriers, more firms must become relentless innovators, embedding constant innovation as a business discipline.

Why is innovation so important now? Many of the competitive differentiators that businesses have enjoyed changed dramatically in the last decade, driven by the rise of developing economies in BRIC states (Brazil, Russia, India, and China) and falling trade barriers. When the United States, Western Europe, and Japan were the innovators and BRIC were the low-cost producers, innovation was only *occasionally* necessary. As industries and economies in BRIC matured, they became far more adept at developing new products and services. Since Western industries had already ceded low-cost production to overseas markets, their ability to add value was primarily in branding and in product design and development. Increasingly, these factors will come under attack as well, by foreign industries and economies that move at a faster pace, with more financial discipline than our older, slower business behemoths.

Additionally, customer demand has accelerated. Product life cycles have shortened dramatically and consumers have become accustomed to new products and services that are updated on a much more frequent basis than was true in the past. Now, consumers have more options in more physical and online outlets than ever before, raising their expectations about the availability and frequency of new products and services. Few firms can rest on their laurels; even recently celebrated industry leaders like Nokia or Dell can be quickly supplanted if they don't consistently meet or exceed customer demands for new products and services.

If larger firms in Western economies don't relearn innovation and reintroduce it to their business models, they'll have little competitive advantage left. These companies will be reduced to distributing products created by faster moving and more innovative competitors or end up attempting to acquire these competitors.

I'm making a distinction that I refer to throughout this book: innovation must become a consistent capability that's sustained over time. Discrete, "one time" innovation initiatives are rarely fruitful. Of course, any firm, in any industry, can innovate once. For proof, look no further than the fact that your firm, regardless of its age or stage, was brought into being from nothing and it had to offer a compelling product or service that was differentiated from its competition. As the firm grew and matured, the innovation focus may have withered while continuity and predictability were consistently reinforced. My aim in this book is to demonstrate that such predictability will work against your firm as innovation grows in importance, while those organizations that choose to develop a sustainable business model with innovation at the core will experience great success.

The Benefits of Innovation and Efficiency

There's a significant amount of work to be done if your firm makes the decision to follow in the path of other relentless innovators. It won't happen overnight and it won't be easy, but the benefits your firm will realize will make the investment worth the effort. The benefits include

- Becoming far more proactive rather than reactive
- Eliminating a lot of firefighting
- Causing other firms to react to your new product, service, and business model introductions
- Employees who are more engaged and who use a broader set of tools and techniques to accomplish strategic goals
- Deeper capabilities to define and achieve strategic goals
- Increased revenues and profits while retaining efficient cost management
- Sustained market differentiation and favorable media and press coverage
- Increased ability to leverage internal knowledge and external partnerships

It is my goal for this book to help your team achieve these results through relentless innovation. Any one of these benefits would be welcome in most corporations, but relentless innovators—firms that balance their investments in efficiency and innovation—can obtain *all* of them.

RELENTLESS
RELENTLESS
RELENTLESS
RELENTLESS
RELENTLESS
INNOVATION

Chapter 1

The Mythology of Innovation

In this chapter, we examine the myths that surround innovation and the misconceptions that arise based on these myths. These myths create overly simplistic explanations for innovation success, and they cause individuals and firms to place importance on factors that don't significantly contribute to innovation success and to overlook other more important factors. Confusion and misunderstandings about the role of innovation, its importance, and its ability to truly affect a firm's success, can, and will, hurt your company.

To begin, let's first examine some easily identifiable successful innovators that receive a great deal of attention, and identify the myths about their innovation success. Firms like Apple, Google, 3M, Target, and Procter & Gamble (P&G) usually leap to mind. These firms have demonstrated an ability to create new products and services that establish new markets or delight customers, while innovating on a relatively consistent basis. Such companies receive heavy media coverage and publicity because of their unusual innovation focus and accomplishments. Along with a number of other firms, these firms comprise a handful of consistent, relentless innovators.

In the United States alone there are hundreds of large, successful firms with recognizable brand names that we encounter every day. We constantly hear innovation success stories about firms like Apple and Procter & Gamble, but we rarely hear about innovation in their direct competitors, Dell and Unilever, much less about innovation in any of the thousands of firms worldwide that compete in these markets. In every region and industry the same pattern is repeated: a small handful of firms are recognized as consistent innovators, used as case studies and examples, while we hear little or nothing about innovation in the vast majority of the other firms in those industries.

So what is it that differentiates a successful, consistent innovator from its close competitors, firms of the same relative size that compete in the same industries and geographies, that aren't viewed as innovative? What factors or attributes accelerate innovation in these successful companies? Are those factors or attributes lacking or underrepresented in lower performing firms? Or are firms like Apple and Google better at attracting marketing and publicity? Is it safe to say that the majority of firms in every region of the globe are not innovative, or is it simply that they don't receive as much media attention? What happens at Target that does not take place at Kmart? What is Apple doing that Dell is not? And what about 3M compared to Avery Dennison?

Several possible factors spring to mind, including the executive management, the nature of the industry, or the capabilities of a firm's research and development teams. Much of the mythology built around innovation identifies these factors as the main components of innovation success and it is true that each of them may contribute to a stronger innovation capability. But in the long run, *none of them are the key drivers*. Let's review the myths and debunk the conventional wisdom, then confront the simpler realities.

Executive Management

Myth: Individual, innovative leadership accounts for the majority of a firm's success.
Truth: Sustained innovation success does not rely on visionary leaders alone.

In the 1990s, a cult of personality arose around some senior executives, especially individuals like Jack Welch of General Electric and Lou Gerstner of IBM. The media led the public to believe that these CEOs accounted for much of their firms' success while they were at the helm. During Welch's tenure at GE he implemented several programs that were attributed with driving new value and differentiation for the company, including ranking employees into categories and only participating in markets or industries in which GE could be one of the top three players. Many analysts have also attributed much of GE's success in the 1980s and 1990s to Welch's leadership.

Strong, visionary leaders matter, but do visionary leaders account for the differences in innovation competence? Certainly, to some degree. For example, everyone recognizes Steve Jobs's influence on Apple and the company's decade-long dominance in consumer electronics and innovation. Jobs, however, isn't the only visionary leader in the computing space, which was created by a number of innovative trend-setters.

Look no further than Kenneth Olson, the founder of Digital Equipment Corporation (DEC), who disrupted the mainframe market with minicomputers, but failed to see the further disruption of the minicomputer market by the personal computer. He is attributed as saying "there is no reason anyone would want a computer in their home."[1] Although he was a visionary leader, Olson did not foresee the imminent changes in the computing market, and DEC was soon disrupted by personal computer (PC) manufacturers such as Compaq, which made the first "portable" PC.

Michael Dell at Dell Computer is every bit as dynamic a leader as Jobs is at Apple, and he was heralded as an innovative leader in the 1990s, constantly on the cover of magazines like *Fortune* and *Forbes*. Dell disrupted the existing business model in the PC market, which enabled his company to grow faster and supplant many larger and well-established firms, including Compaq. In fact, far more people own Dell PCs than own Apple PCs, yet Jobs is constantly feted as an innovator while Dell is hardly considered in the same league.

Dell and Olson were both recognized for their vision and innovative capabilities at a point in time, *but their firms did not sustain innovation over time*. But, back to the initial question of how much impact a CEO has on innovation. If we assert that Jobs is a unique case, can we identify innovative firms that don't have visionary CEOs? Certainly; W. L. Gore is an excellent example.

W. L. Gore is a privately held firm with more than $2.5 billion in revenue, headquartered in Newark, Delaware. Gore manufactures Gore-Tex, the waterproof, breathable fabric that is used in a wide range of outdoor clothing and gear. The company has sought and found numerous uses for its PFTE polymer, creating dental floss, coatings for guitar strings, medical devices, and other applications. Beyond product innovation, however, Gore is also an innovator in organizational structure. Gore has an exceptionally flat organizational structure with no formal reporting hierarchies or organizational charts—its CEO was actually elected by its employees. Innovation at the company is therefore driven not by a single visionary CEO, but by the individuals and teams throughout the business.

Further, consider Target or 3M, firms identified earlier, which are far more innovative than their competitors. While these firms are recognized as innovation leaders, I suspect most people would have difficulty picking out *any* member of the executive team of either firm in a police lineup.

Another thought experiment may help clarify whether or not executive leadership is a significant driver or barrier for innovation. Let's assume that Steve Jobs could be magically and instantly transported to Austin, Texas, where he becomes the CEO of Dell. If this were to happen, do you think Dell would become dramatically more innovative overnight, or even in several years? If Target's CEO was recruited to Kmart, or 3M's CEO was remanded to become the CEO of an abrasives company, would those firms instantly become innovative? Would these firms attain the level of relentless innovation of the leaders in their industries or markets, even over time?

> While engaged leadership is important for innovation success, sustained innovation can't rely on one leader; it must be a cultural phenomenon.

I'd stipulate that the answer is no. Simply put, there's more to sustained innovation than a visionary executive. Visionary, innovative, executive leadership may occur periodically, and while it may contribute to sustained innovation, it is not the only contributor to successful, long-term innovation. Sustained innovation success does not rely on visionary leaders alone.

Industry Competition and Specifics

Myth: The level of industry competition dictates the amount of innovation.
Truth: Industry competition is a factor in fostering innovation, but it doesn't guarantee innovation leadership.

If executive leadership alone doesn't account for innovation success, then perhaps the level of industry competition fosters more innovation. After all, it seems some industries are more innovative than others. A look at the mobile phone handset market provides perspective on a highly competitive and innovative industry. Consumers expect their wireless devices to offer valuable new features and capabilities. Yet, recent history suggests that while many firms in the space have been considered innovative, few of them have sustained leadership for any length of time. Nokia is a great case study in this regard as it was considered the market leader in innovative handsets for many years.

Nokia is an example of a company that has reinvented itself as times and needs changed. Originally a paper company, the firm has shifted its focus and business model at least three times over the course of almost 150 years. Nokia entered the cellular handset market in the late 1980s and as of 2010 was the leading handset manufacturer in terms of volume.[2] Yet its market share has dropped precipitously according to industry analysts[3] as it has failed to anticipate new needs and offer compelling new products.

At the time Nokia was the leading handset developer, its researchers actually designed a touchscreen mobile handset (this was years before Apple's iPhone), but the concept was rejected by executive management, which had become complacent and comfortable with current profits.[4] In early 2011, Nokia's CEO wrote an open letter to his employees, describing Nokia's position in the handset space as a "burning platform" based on the company's shrinking market share.[5]

As Nokia stumbled, Motorola took its place as the innovation leader in the handset industry with the RAZR phone, for a short period. The designers of the RAZR were featured in the business press and were hailed as the new leaders in cell phone design. Yet in just a few years Motorola was dethroned by Apple, showing that it was no more able to innovate consistently over time in the cell phone space than Nokia. It remains to be seen whether Apple will suffer a similar fate with the introduction of the Android operating system and new smartphones based on that technology.

> While the competitive nature of an industry does increase the likelihood of innovation, it does not guarantee a firm will sustain innovation focus.

The point is that within less than a decade several firms wore the crown as the "innovation" leader in cell phone/smartphone development and design, and all of them demonstrated periodic innovation. Yet only Apple appears to be able to sustain innovation. Just because one firm held the leadership mantle and received higher profits during its own leadership period has not meant that such firms could *sustain* innovation over time.

The Fast Follower

Myth: It is possible for firms to copy the product or service offerings of market leaders while retaining competitive advantage through low costs or higher service.
Truth: To remain competitive, firms must increase their innovation capabilities instead of playing "follow the leader."

A quick review of firms in the United States demonstrates that most industries or markets have one well-established innovator and several "fast followers." The majority of firms in any industry don't heavily invest in innovation. Most companies assume they can copy the strategies of the leader in their market and still retain competitive advantage through low cost or higher service—or simply through the lethargy of their customer base. Such organizations will even argue that their strategy is to be a "fast follower." This strategy, however, is usually a difficult one to pursue and it is increasingly a dangerous proposition. There are at least four problems with a business plan of this kind.

The first problem is in the word "fast." Customer demand and expectations are changing much more quickly than many firms have the ability to keep up with. Few products or services have the luxury of extended life cycles or little competition. A growing base of consumers with new expectations and new demands only fuels the fire for more products and services. Firms that claim to be fast followers are often merely just followers. As a firm grows and matures, its bureaucracy, decisions, and approvals inhibit its ability to bring a new product to market quickly. The company can't respond fast enough to innovators or consumer demands. In this period of rapid change and global competition, innovation isn't a "nice to have" but an important core competence; those firms that can't keep up will inevitably perish.

The second problem with "fast" followers is that they become accustomed to *following*. Since these companies don't exercise any creativity or innovation skills, those capabilities have atrophied or they aren't valued within their organizations. This lack of innovation skills leaves the fast follower with only one recourse: to eliminate costs and inefficiencies since they can't hope to command the attention and margins that accrue to innovators. Given new economic shifts, global competition, and customer demand, firms that cannot create new, interesting products and services exist on the very brink. To remain competitive, firms that haven't relied on innovation as an advantage must increase their innovation capabilities, not try replicating others' successes.

Third, "fast" followers often don't understand what features or benefits the customer values in a product, and what challenges or issues exist in those products. By simply copying an existing product or service, they risk duplicating all the problems or issues that exist within the innovative product. Since the "fast" follower does little research, the company often doesn't know which features or benefits are important and should be emphasized, or what hidden issues or concerns exist with the product. "Fast" followers often make the same mistakes as innovators do, but they have less opportunity to respond and encounter a customer base that has recognized both the benefits of the product or service and the issues or constraints.

Finally, "fast" followers suffer the most as new innovations enter a market. They are more accustomed to implementing the business models and offerings of the innovation leaders after the models have been proven. Fresh entrants, unbound by the shape and structure of the market or competition, will enter to disrupt the existing order and make older products, services, and companies obsolete. Innovators by their very nature are constantly scanning the horizon, looking for emerging threats and new entrants. They spot disruptive trends and shift nimbly into new opportunities. Industry laggards and fast followers are impacted by disruptions far more than innovators, but the impact is more severe on fast followers since laggards really had little to lose. Since such companies are neither fast nor particularly insightful, they lose the most in a market disruption as they can't shift away from their existing models and structures quickly enough.

The "fast follower" strategy is increasingly a difficult business proposition. Firms that focus their efforts on innovation rather than fast duplication will succeed.

As Michael Treacy established in his book *The Discipline of Market Leaders*, there are three differentiated positions in any market: product leadership, operational excellence, and customer intimacy. Innovation is a tool that can help an organization achieve leadership in any one of these differentiated pursuits, but clearly only one firm in an industry can be the "best" at any of these strategies. For example, we could argue that in the retail space, Target is the product leader, partnering with leading designers to bring interesting, attractive, and affordable products to the mass market. Wal-Mart is the operational efficiency leader, innovating new data streams and distribution tactics to keep costs and prices low. Nordstrom is the customer intimacy leader, creating a completely unique and valuable relationship with its customers. Every other retail firm lags behind these firms in one or more of the three strategic areas, and new competitors seek to enter the retail space and disrupt the leaders, much less the laggards.

Sustained Innovation

Myth: Due to changes in a globalizing world, no firm can sustain innovation leadership over the long term.
Truth: Sustained innovation resides in factors that companies can control.

Some observers argue that given heightened competition, accelerating global trade, and increasing customer demands, no firm can sustain innovation leadership over the long term. This argument, however, ignores the results of a firm like 3M.

Except for a brief period between 2000 and 2005 under former GE executive James McNerney, whose focus was on profitability and efficiency, 3M has had a long history of innovation leadership, creating a range of products and services. Certainly the Post-It is probably the most well known, but over the last 50 years 3M has entered countless markets and industries, tailoring new innovations to different geographies, technologies, and market needs. Though 3M continued to innovate in spite of McNerney's focus on efficiency, when George Buckley replaced him as CEO, one of Buckley's first actions was to *reemphasize* innovation as a core capability, providing fresh focus and funding for those activities.

Innovation is a cultural phenomenon which can be enhanced or inhibited by leaders, culture, and strategy.

In my experience, it is completely possible for a firm to develop and sustain an innovation capability over time, just as a firm is able to create and sustain market leadership over time. Innovation capability resides less in markets, strategies, technologies, or leadership than we typically suppose, and more specifically in factors that companies *can* control—culture, business attitudes and perspectives, focus, and intent. That's the real lesson we can learn from relentless innovators: what drives long-term, successful innovation are the same factors that shape the way people think and act in any business: operating models, strategies, rewards, culture, and processes.

Innovation "Business as Usual"

If visionary leaders, competitive markets, or complex innovation techniques don't sustain innovation in an organization over time, what does? In the work OVO has done we've found that the firms that have the greatest success innovating over time, regardless of circumstance, market conditions, leadership, or customer demand, have successfully developed a concept I'll call *innovation business as usual.* By this I mean that innovation is "business as usual," a normal operation, in these successful organizations.

> The most important factor sustaining innovation in leading innovative firms is an operating model that considers innovation as "business as usual."

Further, an innovation business as usual approach is sustained by the most important team members: middle management. In most firms, efficiency and effectiveness, and a focus on short-term predictability and profitability are "business as usual" while innovation is at best episodic based on reactions to external events. Let's take a look in the following chapter at how business as usual and middle management connect to create a barrier against innovation business as usual.

Chapter 2

The Real Barriers to Innovation: Business as Usual and Middle Management

In the previous chapter, I examined and rejected a number of the "myths" that surround innovation, that make innovation seem more complex and difficult than it needs to be. In this chapter, I'll examine the two significant barriers to innovation, which, if properly engaged, can become two significant *drivers* for innovation success. The first barrier, business as usual, defines how work is done. The second barrier, middle management, defines what work gets done. I'll examine why these two factors are barriers to innovation, and how they can become enablers.

Business as usual (BAU) can be defined as the expectations, attitudes, processes, and methods that a business follows to get work done. BAU may be formal, carefully documented over time, or it may represent informal, generally accepted norms. You might be tempted to lump all these factors into a bucket called "culture," but the issue is larger than that. In addition to being the aggregate of all the written and unwritten rules, methods, processes, attitudes, expectations, and perspectives about how the business should run, business as usual *incorporates* corporate culture and communication, corporate history, attitudes about risk, ambiguity and uncertainty, evaluation, compensation, and how people are punished or rewarded for the outcomes they generate. Though formal agreements and workflow are major parts of an organization's BAU, the informal hierarchies, decisions, shortcuts, and workarounds are just as important. The vast majority of businesses in every geography and industry operate in this way.

Once fully adopted, BAU describes how work is performed most effectively, and bends everything to its will to ensure consistent compliance throughout the organization. Business as usual is developed, supported, and reinforced by a cadre of managers called "middle management."

Middle management (MM) encompasses anyone who isn't a "front line" employee dealing with specific and discrete tasks or anyone who isn't a senior executive. In this definition, middle managers are the people tasked with the job of achieving quarterly results, in terms of profit and costs, and in units of production and customer satisfaction. These middle managers achieve their goals and the goals of the organization, by reinforcing the common business-as-usual operating model.

Importance to Innovation

Business as usual and middle management may seem unimportant from an innovation perspective. After all, the public is constantly bombarded with books, articles, and presentations that claim to provide the "secrets" of innovation success, but to my knowledge, none of them focus on anything as "pedestrian" as middle management and business as usual. As discussed previously, most of the literature on the topic of innovation focuses on visionary leaders, competitive markets, or specific innovation techniques. While those factors may improve the probability of successful innovation, they mean little if middle managers aren't invested in innovation and if the BAU operating model rejects innovative ideas.

What's more, business as usual and middle managers are intertwined, mutually reinforcing, and almost inseparable. BAU is constantly reinforced by middle managers who are responsible to ensure that a firm is run effectively and achieves its goals. Middle managers are alert to any initiatives or threats that disrupt or distract BAU and that will undermine quarterly results (the yardstick and ultimately the reward structure of the organization).

> Business as usual and middle management may seem mundane, but these two factors will dictate whether or not your firm can innovate successfully over time.

While BAU and MM may seem mundane, in my experience they determine whether or not a firm can innovate successfully over time or if innovation is a "flash in the pan" that is not sustained. Any company with the appropriate amount of market pressure, serendipity, or insight, can innovate once, or perhaps for a short period of time. What we, as the business community, need to focus on intently is what allows firms like 3M, or Procter & Gamble, or W. L. Gore, all relentless innovators, to innovate consistently over a long period of time, while their competitors do not. As discussed, the key differences between these two separate types of companies are not visionary leadership, markets, insights, technologies, or funding, but in the much more "mundane" factors of business as usual and middle management. Let's deconstruct those two factors briefly to understand why.

Business as Usual

Every firm has formal methods, processes, and hierarchies along with unwritten, informal but generally accepted agreements about how work gets done effectively, known as their "operating models." Everyone implicitly understands how work is accomplished, what initiatives receive priority, where funding and staffing will flow, and what is necessary to achieve the goals set forth for the firm.

Typically, the longer a firm is in business the more definitive the BAU operating model becomes, permeating the thought process and approach of the majority of employees. Most new or entrepreneurial firms spend a lot of time "redefining the wheel," establishing their operating model through trial and error. As a firm matures, the historical approaches, decisions, and formal and informal processes become codified as "the way we do things." Mature firms leverage BAU models almost instinctively, even in the face of well-documented process maps and workflow that may speak against what has become the company's status quo.

As the business as usual model develops, it acquires supporters who understand the value it creates. Older, established employees rely on the model for consistency and optimal effectiveness. Newly hired employees quickly learn the formal and informal rules and methods, and grow to accept both the benefits and constraints of the approach. Over time this operating model becomes the de facto methodology. Every consideration is taken to ensure that the operating model is efficient, effective, and, most importantly, well understood. Nothing is allowed to threaten the operating model or distract it from achieving quarterly results.

The Problem with BAU

An engrained BAU is powerful enough to cause people to accept poor decisions, unwieldy processes, poor communication, contradictory reward structures, and the absence of clear goals *without question.* Intelligent workers are forced to do unintelligent tasks, or create ad hoc workarounds that become part of the accepted practice, rather than addressing the original shortcoming or problem. In my experience, BAU has consistently been recognized as an issue, though it becomes an easy rationale to avoid new initiatives. I've participated in meetings in which my clients have debated interesting, valuable ideas and rejected them out of hand because of the difficulties of implementation in the face of these operating models.

Managers and employees accept this framework because they recognize that a commonly reinforced operating model allows the business to operate in the background, as second nature, rather than demand the full attention of the staff. The more predictable this engine of business becomes, the less management it requires. As predictability and efficiency of the internal processes increase, greater defenses are put in place to sustain existing practices.

> Ever-increasing focus on efficiency creates an innovation trap: the more efficient BAU becomes, the more the firm seeks to protect and isolate BAU, leading to less and less innovation.

Other management initiatives have also enhanced the prominence of business as usual. As outsourcing and downsizing increase, the remaining management team is ever more dependent on a self-reinforcing process operating with little oversight. This dependency spawns a vicious circle. A closely defined operating model that is carefully followed allows more productivity and efficiency, and therefore becomes more important to protect. The more efficient the model, the more important it is to protect it from distraction and disruption. Over time, broken or misaligned processes, inadequate information, contradictory demands, and other seemingly disruptive factors become part of the status quo, accepted, managed, and overcome, rather than challenged, improved, and accelerated. Business as usual becomes entrenched.

Of course, it could be argued that even if the BAU process and methods aren't optimal, it is far better for every employee to adhere to one common method, rather than to establish a number of more efficient but competing methods and standards. In fact, that situation is exactly what the main management fad of the 1990s introduced: documented complacency. The International Organization for Standardization (ISO) certification that was the height of the Total Quality Management (TQM) and Business Process Reengineering (BPR) movement doesn't claim that the processes as defined are the best or most effective processes, *only that the processes are documented and people understand them and adhere to them.* Thus, business is accomplished in a relatively effective way, with one common approach that is reinforced by the employees using written and unwritten rules to sustain the capability, whether the process is optimal or not.

After BAU is the accepted practice, it becomes familiar and increasingly difficult to alter or replace. Individuals who suggest changes to the model or introduce initiatives to change the business model are rejected or derailed. Like any bureaucracy, business as usual also develops defenses to sustain its existence, supported by those people who rely on it for its lack of risk and predictability. These people have a stake in sustaining a common, consistent operating model to achieve results repeatedly.

The more front line employees, middle management, and executives who accept the operating model and find it beneficial, the more valuable the model becomes, creating a network effect. The organization is ever more reliant on the model for success and any efforts to change it grow ever more difficult. To understand the main reason for why this situation occurs, let's look at BAU's chief supporters: middle management.

Middle Management

Middle management has been greatly derided over the last 20 years of strategic management thinking. Many strategic thinkers and academics argue that middle managers serve little purpose in an age where there is more need for rapid response to customers and markets and less need for hierarchy—especially in an era when the average worker is exceptionally well educated and capable of making decisions and taking action for herself.

The function of middle management has changed little over time: its role is to enact the vision and strategy of the firm, defined by senior executives, as efficiently and effectively as possible, while ensuring that quarterly and annual results match or exceed what executives promise to Wall Street. MM deploys staff, funding, and other resources to accomplish these goals. Ensuring efficient, effective business processes and achieving quarterly results in an environment that is constantly evolving and where customer demands are increasing, places an increased emphasis on the established and consistent operating model. When so many factors are uncertain and demands for returns constantly increase, even though resources are held flat or decline, middle managers need a reliable way to do business that isn't subject to change, that is easy to learn and scale, and that is relatively predictable. Middle managers therefore become the guardians of a BAU process. In an ever-changing world, the model remains predictable, easy to understand, and it usually produces the *necessary* results.

> In a world where so many factors are in flux, middle managers count on business as usual as a reliable, trustworthy way to get work done efficiently and effectively and they are therefore avid defenders of the model, often rejecting innovation.

The Conflict

While middle managers seek models that are easily understood and consistent, markets, customers, and technologies aren't stagnant. Products age and become obsolete. New technologies, business models, and services arise. New firms enter markets, disrupt the status quo, and dramatically change customer expectations and demands. All of these factors introduce even more complexity in the business, which middle managers struggle to control.

The only answer to many of these challenges is innovation: the creation of new products, services, and business models that allow firms either to compete in a changing market or to establish new customer segments or markets. But while innovation has such significant benefits, it introduces a major threat to the "sacred" core of the business.

Innovation versus BAU

Innovation disrupts the business-as-usual operating model that aims to avoid risk, change, and uncertainty, so innovation and BAU seldom exist comfortably side by side. Innovation requires unfamiliar tools and demands new customer insights. Firms must create radically new products and services, many of which may not be successful in the marketplace since they require new relationships and partnerships that are unfamiliar. Achieving this goal can be incredibly difficult, especially when trying to attract prospects who may not have been customers previously. Innovation may also require changes to organizational structure, while cannibalizing existing products or forcing changes to existing business models.

Simply put: *everything that's necessary for innovation to succeed threatens the existence and sustainability of the long-developed BAU process.* Innovation attempts to place the "round peg" of new ideas into the "square hole" of existing business-as-usual processes and expectations. Clearly, new, nascent ideas are the losers in that confrontation, because the BAU model has been validated and reinforced over time.

> Innovation often fails when firms attempt to manage new, radical ideas in traditional, business-as-usual processes. This problem is a "square peg, round hole" one; the new ideas will be rejected, not the existing processes.

Stuck in the Middle

As defenders and supporters of the business-as-usual process, middle managers are caught in an exceptionally awkward position. Innovative initiatives are usually dictated by executives who want more organic growth or greater differentiation. MM enacts the strategic goals and directions of the executive team. Yet no one understands more clearly than middle management the impact of an innovation initiative on a BAU model and the threat it poses to its existence. In the face of that threat, middle managers must make critical decisions to determine if a focus on innovation is one of these:

- A momentary whim, to be acknowledged and then ignored
- An important but discrete project or initiative
- A new way of business life

Middle managers must quickly decide if an innovation request is merely the "flavor of the day." If that is the case, they can safely ignore the innovation, thus sustaining and protecting the BAU process. What is much more difficult is determining how to accomplish two mutually exclusive tasks: running an effective, efficient BAU model, at the same time running a disruptive innovation effort that will require funds and resources originally intended to sustain the company's operations. Additionally, the middle manager will have to decide if, while introducing an innovation element into BAU, the company can continue to achieve the anticipated results that depend on the BAU process working effectively.

It is the conflicting nature of these two tasks—sustaining a highly efficient business that meets quarterly objectives on one hand and innovating to drive new products and services on the other—that often forces middle managers to make a hasty, though critical, decision. The emphasis on efficiency, effectiveness, and achieving quarterly financial goals has far outstripped the importance of innovation in most organizations and firms, so middle managers have repeatedly emphasized efficiency while downplaying innovation initiatives.

The New Operating Model

Relentless innovators respond to the conundrum previously identified by creating a new business-as-usual operating model, introduced in the last chapter: the *innovation business-as-usual operating model.* Relentless innovators have created an effective balance between innovation and efficiency in their operating models, demonstrated by their priorities, communications, and processes. They have built this concept into their decision making, cultural biases, and attitudes, supporting the realization that a firm must innovate consistently to thrive in today's marketplace.

Rather than viewing innovation as a threat to business as usual, these firms have established processes that assume innovation will happen on a consistent basis, instead of ad hoc on the whim of an executive or in the face of an immediate threat posed by a competitor. In successful, long-term innovative companies like 3M or W. L. Gore, not only is innovation the BAU process, but middle managers are the driving and enabling forces that sustain the process. Clearly, middle managers at these and other innovative firms are under pressure to deliver consistent results as well, which they are able to achieve. What's surprising to discover is that relentless innovators, while constantly creating new products and services, are also efficient in their use of inputs and resources, demonstrating a balance between efficiency and innovation.

Efficiency Metrics

Evaluating the "efficiency" of any publicly traded firm can be difficult. There are, however, several economic ratios investors analyze to determine if a firm uses its assets to generate value above those assets and in comparison to others in its industry. Anyone familiar with investing principles will recognize return on assets (ROA), return on investment (ROI), and return on equity (ROE). After a review of market metrics, and considering the different industries the relentless innovators represent, I chose two other efficiency metrics recognized by investors:

return on invested capital (ROIC) and economic value added (EVA). Both ROIC and EVA indicate how, and to what extent, a firm is managing its assets, capital, investments, and people efficiently. The return on invested capital metric describes just that, the returns a company generates on its investments. The EVA is an estimate of a firm's profits after the return to its shareholders. (ROIC and EVA are not publicly available for private firms, so W. L. Gore will not be a part of the following analysis.)

Using ROIC and EVA, we can assert that Apple, 3M, Google, and Procter & Gamble are at least as efficient in their use of capital as their competitors. The ROIC measures for these relentless innovators are at least as efficient as their close competitors, if not more so. Apple's ROIC is almost twice that of Dell's, while the same is true for 3M when compared to Avery Dennison. From a return on invested capital perspective, Procter & Gamble is at par with its closest competitors Johnson & Johnson and Kimberly-Clark. If we examine economic value added as a yardstick, Apple maintained an EVA momentum more than *80 times higher* than Dell between 2008 and 2011, according to an article on the Motley Fool Web site.[1] 3M leads its competitors on an EVA basis and Procter & Gamble is at par with its competitors. At least for this handful of firms, innovation and organizational efficiency comfortably coexist. They demonstrate that an "either/or" decision between efficiency and innovation isn't mutually exclusive.

The authors of *The Innovator's DNA* created another interesting metric for determining whether or not firms that claim to be innovative actually are innovative. They asked HOLT, a division of Credit Suisse, to analyze the financial returns of large companies to ascertain the percentage of a firm's market value that is derived from cash flows from existing products and services. If the market value of a firm is higher than existing cash flows, the authors suggest that the company is demonstrating an innovation premium.[2] How do the firms I've called relentless innovators fare under this assessment? Apple, Google, and Procter & Gamble are each ranked in the top 25 of the world's most innovative companies using the authors' methodology.[3] Gore is not listed because it is a private company, and 3M fell just outside of the top 100 firms.

These metrics demonstrate that relentless innovators are at least as efficient as their direct competitors and create market valuation premiums far in excess of their competitors.

Relentless innovators use their people and assets efficiently and drive higher market valuations based on the expectation of innovation success. Their operating models balance innovation and efficiency, delivering outsized returns.

Barriers Can Become Accelerators

These examples help illustrate that the BAU process *can* support innovation business as usual, and middle managers who sustain such a model are those able to become "multidimensional." They are capable of managing an existing business process efficiently *while* managing a consistent innovation effort.

So, though MM and BAU typically clash with innovation, relentless innovators actually *depend on middle managers for sustained innovation success.* An entirely different set of attitudes, behaviors, expectations, and processes exist in relentless innovators as opposed to the vast majority of other firms. The two main innovation barriers discussed throughout this chapter can actually accelerate innovation success if innovation becomes BAU.

The middle managers in relentless innovators are subject to the same expectations as managers in other firms, including quarterly expectations for revenue and profit and the need to run effective, efficient organizations. However, they are also expected to create and sustain processes, business cultures, and attitudes *that welcome, embrace, and accelerate innovation*, rather than seek to delay, derail, and slow it. In the end, what leads to success for relentless innovators is how *unusual* these firms are in their expectation of innovation as a consistent business discipline, coupled with their multidimensional middle managers.

The End of the Beginning

Innovation can become systemic in any organization that decides it is valuable and important. In fact, it's hard to imagine a firm that doesn't require innovation to thrive over the long term. Executive and middle management must shift their thinking, yes, but any firm in any industry can develop more innovative capabilities and sustain innovation over time. *Innovation is a strategic choice rather than an act of fate.*

> Since innovation success is based on strong capabilities that already exist in your business, innovation is a strategic choice rather than an act of fate.

Rethinking business as usual is necessary if your firm is to evolve. While this shift in thought won't be easy, and it will heavily impact middle management, it can be accomplished and it is necessary for your firm to succeed in the long run. Without a significant change to business as usual and the middle managers who support and sustain this approach, no firm can hope to do more than occasional innovation based on dire threats or serendipitous insights. This lack of regular innovation will lead to less interesting products and services, fewer profits reinvested in the business, and even more focus and urgency around the BAU process.

Unfortunately, rejecting or delaying innovation can become a vicious cycle. Continual cost cutting starves the business of new revenues, reduces the number of knowledgeable people, increases the reliance on fewer existing products and revenue streams, and further emphasizes the importance of the BAU operating model. Constant cutting and efficiency improvements only make middle managers and the corporate culture less willing to change, especially in an uncertain and risky direction.

Firms also need to nurture and develop "multidimensional" middle managers who have expertise in efficiency and effectiveness *and* who understand how to support and enable innovation. These managers must implement a new innovation BAU and support that model over time. Innovation must be embraced as business as usual.

So, you and your firm have a choice. You can sustain a safe, comfortable BAU approach as your products and services come under attack, sacrificing profits, which will result in fewer resources being available to reinvest in new products and services. Right sizing and cost cutting will also result in fewer fresh ideas and a heightened focus on existing processes, increasing the difficulty to introduce new products and services. Or you can refocus and repurpose your business-as-usual process to embrace innovation and sustain it over the long run. The first choice is a vicious circle that leads to heightened competition, lower margins, and eventual obsolescence. The second choice leads to greater innovation, higher profits, enhanced differentiation, and long-term success.

As Roger Martin suggested in his book *The Opposable Mind*, it's not an "either-or" proposition, but a "both-and." Firms need sustainable innovation processes in parallel with processes that generate near-term revenue and profits in line with expectations. The first creates differentiation and growth for the future while the latter generates the near-term revenues and profits that sustain the firm in the short run. Business as usual cannot neglect or slight either need, *so the processes must support both.* Simultaneously, the middle managers who sustain these processes must receive training and support to transition to a new innovation business-as-usual capability.

Chapter 3

How Things Work Today

Most people have become accustomed to the concepts of corporate cultures, behaviors, attitudes, and perspectives as beneficial forces that help shape and form the way companies operate. This "operating model" is the shared understanding of how things get done that propels your organization, helps it achieve results, and ensures that resources, people, and inputs are consumed effectively. While BAU is beneficial—a reliable, trusted, universally accepted way of doing business—it is also a sleeping tiger, becoming a man-eater when those methods and processes are threatened. And as we've seen, innovation threatens BAU more than almost any other initiative.

In most situations, the operating model is not a man-eating tiger, it's the "tiger in your tank," to quote an old Exxon commercial, propelling the firm to great results when processes are optimized. We've discussed how beneficial and powerful the model can be. But as any wild animal trainer can tell you, you can take the animal out of the jungle, but you can't take the jungle out of the animal. Also as discussed, business as usual, like any self-respecting bureaucracy, establishes defenses to sustain itself when under attack. These defenses reinforce the bureaucracy and fend off change and attacks on the existing culture. Any initiative that seeks to alter the operating model or that will indirectly impact it is viewed suspiciously—a powerful operating model will resist change aggressively. In fact, the more competent and capable your operating model, the more efficient and effective it is, the more difficult it will be to introduce innovation.

Most executives understand how strong and reactive the operating model can be, and they mistakenly believe they've tamed the beast and put it to use for their benefit. However, there lurks in most corporate bureaucracies the heart of a tiger, willing to defend its turf when under attack. Even the people constantly called on to drive more effectiveness and efficiency, middle management, understand that small changes to long-defined operating models can cause a major disruption at a firm. Continuing our tiger analogy, middle managers are the professional tiger tamers who keep the tiger (the operating model) performing at top efficiency. They are also the ones who understand just how dangerous a perceived attack on the BAU operating model can be. Further, middle managers know they'll bear the burden of any suggested change to the model and they will have to clean up any mistakes. Therefore, *middle managers have more at stake than anyone else when the operating model comes under attack*, and they are the employees most likely to rush to its defense. So it's no wonder that middle managers will

- Rush to BAU's defense
- Seek to sustain *existing* cultures, attitudes, and beliefs whenever possible
- Reject anything that disrupts or threatens the existing model

> Innovation is an insidious threat to business as usual. And like any bureaucracy, the operating model will defend itself vigorously.

Innovation is therefore an unexpected yet insidious threat. Innovation promises great benefits, but it has a subversive nature, demanding changes to trusted, proven methods, perspectives, and processes. Business as usual will resist change, and the more effective the operating model, the greater the resistance.

Well, How Did We Get Here?

How did it come to this? How do so many firms, which get their start as innovators, mature into organizations that view innovation as a significant threat to their businesses? Inertia and complacency emerge as a business grows and matures. It is difficult for any firm to maintain the rapid growth and focus on innovation that nurtured it during the period of growth and expansion. After an all too brief entrepreneurial period, when a firm first springs into being, relying heavily on innovation for success, most firms shift rapidly into a sustaining mode, protecting existing products and services. At this time, they also begin seeking to reduce risk and uncertainty as much as possible.

For many firms, after the first flowering of innovation leads to a valuable set of products or services, the scope and pace of innovation decelerates until it seems almost vestigial, a capability once important but no longer necessary. Innovation, while useful once in the dim recesses of history, is not practiced or implemented currently. Many businesses have buried somewhere in their corporate histories the stories of the first innovators who created new products and services, but often those individuals seem like strange forebears who aren't aligned to current goals and missions. Innovation capabilities remain a part of the lore of many firms, but those references and stories seem misplaced in an era of high efficiency, cost-cutting, and outsourcing.

Management Philosophies Become Innovation Barriers

Management philosophy plays a role in this negative view of innovation as well. Management doctrine states that firms should scale up, take as much market share as possible, and then generate the maximum amount of profits. Profit generation requires either *growing revenues* or *reducing costs.* Cutting costs is often a much simpler proposition than growing revenues, so innovation takes a back seat to cost reduction in an attempt to sustain profits. Looking at the last 20 years of management science, it's easy to see that the main focus over that time has been on cost management, effectiveness, and efficiency.

In the 1980s U.S. firms were rightly criticized for creating poor quality products, especially in comparison to their Japanese competitors, causing a shift in focus by many American businesses to Total Quality Management (TQM). This focus culminated in the creation of the Malcolm Baldrige National Quality Award. The first award was issued to Motorola, Westinghouse, and Globe Metallurgical in 1988.[1] At first this emphasis was on developing products with higher quality, although the focus eventually extended to improving processes (business process reengineering). As quality improved and processes were rearchitected, another shift in the competitive landscape occurred.

President Bill Clinton signed the North American Free Trade Agreement (NAFTA) in 1993, which reduced trade barriers for the United States, Canada, and Mexico.[2] This bill ushered in a tidal wave of free trade, which in turn meant U.S. companies entered new markets and more firms in more countries began offering products and services to the U.S. consumer. As additional trade barriers fell, and companies in countries like China, Brazil, and India began targeting the U.S. market, U.S. firms faced two choices: become exceptionally more efficient (capitalizing on TQM and the new concepts of Six Sigma and Lean) or outsource any activity to a region or country where the cost basis is lower. Some firms chose to follow both strategies.

Six Sigma is simply an outgrowth of the TQM movement, meant to reduce the number of variations in a manufactured product. As Six Sigma demonstrated the ability to reduce errors and variability on the shop floor, its techniques were applied to business processes and management practices. Lean, on the other hand, is focused on doing the most with the least resources. Both Lean and Six Sigma have been applied to problems that are far removed from their original intent and scope, leading to debates about whether or not these techniques are "innovation" tools.

In addition to Six Sigma and Lean, right-sizing, outsourcing, and other management initiatives and tools have honed corporate operating models to exceptionally high efficiency standards. It's no wonder that innovation is considered a threat; over the last 20 years the focus in most U.S. firms has been on crafting the most efficient, effective, high-quality, low-variation operating model possible. Innovation, typically, is just the reverse: uncertain, highly variable, inefficient, and unpredictable. For years, management teams have implemented programs meant to eliminate many of the factors that innovation introduces. The business of big business is efficiency and predictability, not innovation.

Large businesses, especially, aren't organized to innovate. They are structured for efficiency and to offer existing products and services to current customers. As discussed, these structures are important, ensuring low operating costs and delivering value for customers on a consistent basis. In large organizations, then, we have refined the operating model to the extent that innovation has become a threat to how firms operate, rather than a potential benefit.

In most businesses, the operating model has been refined to the point that innovation is viewed as a threat, rather than a potential benefit.

Resisting Innovation

How resistant is the typical organization to innovation? The gaps may surprise you. CEOs consistently report that innovation is one of their top three priorities. Yet a survey in 2010 found that less than 25 percent of manufacturing firms in the United States created an "innovative" product or service in the last three years.[3]

The gap between the expectations of executives about innovation and the actual innovation work under way is stunning.

The gap between expectations established by CEOs and the actual implementation of new ideas by middle managers as products and services is almost 50 percent. While the 25 percent of manufacturing firms that claim to have created innovative new products may seem appallingly low, consider the services industry, in which less than 8 percent of firms admitted creating a new product or service in the last three years.[4] Remember too that more than 80 percent of businesses in the United States are services oriented. These statistics further indicate that the vast majority of companies in the United States are not innovating frequently, and certainly not consistently or "relentlessly."

Clearly, there's a disconnect somewhere. Executives are calling for innovation. For example, George Buckley, the CEO of 3M, told shareholders at the annual shareholders meeting that "the company's unwavering commitment to innovation is reenergizing new product development and transforming 3M into a stronger company."[5] SAS, a large software developer, begins their 2010 annual report with the statement "Innovation is at the heart of what we do."[6] These are just two examples of CEOs and executives placing emphasis on innovation as an avenue for growth. President Obama made innovation a cornerstone of his 2011 State of the Union address. Books, articles, and experts expound on the value and importance of innovation. Moreover, customers *expect* new products and services, while the demand for new products and services continues to rise as globalization increases and new markets grow in China, India, and other developing countries. Every executive, it seems, wants innovation, yet many organizations, as is apparent from a National Science Foundation survey published in 2010,[7] are avoiding innovation.

The last statement is unfair, actually. Though the majority of firms attempt to maintain the status quo, and therefore do not put a priority on innovation, there are many firms that try implementing innovative ideas and efforts but fail, or find the results less than satisfactory. Some reasons for failure or poor results include

- Poorly communicated strategy
- Lack of resources
- Demands for quarterly results
- Fear of uncertainty and/or risk

But, as discussed, these barriers can be overcome. What stops most innovation efforts most frequently is the formal and informal "operating model" that is so integral to efficient operations.

Reinforcing the BAU Operating Model

There are a number of management tools and strategies that can improve or optimize existing business processes. Some of these techniques, like Six Sigma and Lean, have demonstrated quite powerful results in improving effectiveness and eliminating waste and inefficiency in existing processes. However, it is my stipulation that many of these tools, while improving effectiveness in the short run, also unintentionally reinforce the existing operating model, leaving no room for innovation. This unintentional reinforcement of the BAU processes happens because experts in such management strategies apply their knowledge to improve the *existing* processes, without questioning whether or not those processes are the "right" or optimal ones.

Once any work is done to improve an existing process through management tools, it becomes much more difficult to question whether or not the process is valuable and useful. Let's look at Six Sigma and Lean to understand why they are valuable in the short run, but why they also can be unintentionally disastrous for innovation.

Six Sigma

Six Sigma was introduced as a method to reduce variance and defects and improve quality in manufactured parts. It was originally developed by Motorola in the late 1980s, and it was more widely adopted by U.S. firms in the early 1990s.[8] Its name refers to the number of standard deviations within a normal distribution. Six Sigma, or six standard deviations from the mean, equates to 3.4 defects per million parts. Over time, the name has become synonymous with the highest quality and minimal variability in manufactured parts and processes. As Six Sigma methods demonstrated dramatic improvements in quality on the shop floor, many adherents extended its use into other areas to refine internal processes and methods far from the shop floor.

The methodology introduces small improvements to existing processes to eliminate errors. Rarely, however, does Six Sigma recommend a complete rethinking of a process, decision-making capability, or perspective. So while incremental improvement of BAU occurs, a stamp of approval is given to current processes, decisions, and perspectives. With Six Sigma's deployment across different departments, fields, and industries, it now is often confused for an *innovation* methodology. Instead, Six Sigma is an example of a trusted, effective tool stretched to extreme uses, sometimes far from its original intent. Consider the major differences between Six Sigma and innovation tools and techniques. Six Sigma

- Seeks incremental improvement to existing processes and products, so *at best* it is an *incremental* innovation technique
- Does not seek to disrupt existing processes or introduce new services or business models, so its focus is limited
- Is a completely inwardly focused tool, intent on solving internal problems and challenges

Innovation, however, seeks to solve customer challenges or open new markets, through new products, services, business models, and experiences.

Six Sigma seeks at best a *refinement* of existing processes and products. The method may buttress existing processes that are actually barriers to innovation, rather than seek to significantly change or replace those processes. Further magnifying the problem, Six Sigma was initially introduced by third-party consultants, but over time many organizations have developed deep internal capabilities to deploy Six Sigma tools and techniques. Many of these internal Six Sigma experts, often called "green belts" or "black belts" based on their training and experience, become defenders of the very processes they are meant to improve.

To a great extent, many of the Six Sigma practitioners are faced with the "curse of knowledge": they are so close to the problem and what they see as the solution that they can't grasp the bigger challenge. Their perspectives are often limited to modifying the existing process rather than replacing it or eliminating it. While Six Sigma adherents seek to eliminate inefficiencies, variability, and quality issues from existing processes, Lean adherents seek to perform as much work with as little input as possible.

Lean

Lean techniques originate from the concept of "Just in Time" manufacturing techniques, first deployed in the automotive industry in Japan. These concepts were further developed and documented in the *Sloan Management Review* in an article titled "Triumph of the Lean Manufacturing System" by John Krafcik.[9] This method seeks to reduce to a bare minimum the inputs and resources necessary to create a product or service. Stripping down unnecessary inventory, assets, people, and actions means that a process can use fewer resources to accomplish the same goals, often with lower costs. While Six Sigma reduces variability and errors and increases quality, Lean techniques ensure the same high-quality outputs are delivered with as little raw material, inventory, labor, and other inputs as possible. Lean, then, is almost entirely inwardly focused and concentrates primarily on eliminating cost in the creation of a product or service, rather than innovating the product, business model, or customer experience.

Reinforcing Efficiency over Innovation

While these techniques were first introduced in the 1980s and 1990s, today most organizations have teams of Six Sigma "black belts" or Lean consultants who are constantly refining the existing BAU process. Some firms deploy both concepts to hone their processes and improve efficiency and effectiveness, and many have gained cost savings and increased throughput from the use of these methods. However, those gains have come at the cost of a BAU operating model that's even more embedded, entrenched, and protected than ever before. After all, if a firm spent thousands or even millions of dollars "improving" the business-as-usual process, isn't it likely that it will not welcome initiatives that threaten the efficiency of the process?

> Important management tools like Six Sigma and Lean are reinforcing operating models that resist innovation.

Managers with Lean or Six Sigma credentials have a vested interest in the operating model as it exists, and profit from incremental changes rather than from reconsidering the purpose and goal of the model. As the need for innovation has become evident, many Six Sigma and Lean advocates are attempting to redefine themselves and their tools. This movement is at best limited in its vision and scope, and at worst derailing serious innovation efforts, since internal Six Sigma and Lean teams have thousands of hours and millions of dollars invested in existing operating models. Many of these advocates have become defenders of BAU, rather than using the appropriate tools and techniques to shift the operating model to embrace more innovation. Rather than embracing new methods and tools meant to help a firm innovate, they are using tools meant to create more efficiency to try to accomplish innovation goals. While I've focused primarily on Lean and Six Sigma, they are just *two of many* management approaches that improve operational efficiency while building barriers to innovation.

Where We Are

So, where are we today? The vast majority of CEOs recognize the importance of innovation, seeing its capabilities in firms like 3M, Google, and Apple. Executives therefore want the benefits that innovation can create. They have built and refined, however, an engine of prosperity that has delivered results for years, or even decades, using highly touted techniques and methodologies. Meanwhile, these firms have been through massive restructuring to equip themselves to face increased competition, and they have downsized, right-sized, and outsourced everything that made sense.

Today, these firms are highly optimized to create specific, quality outputs and occasionally incremental products and services—all engineered to achieve quarterly financial goals. These firms operate on the thinnest of margins with more productivity and efficiency than most management theorists believe is possible. They are constructs of operational beauty, but they are not nimble and they cannot adapt to the one truly sustainable competitive advantage: innovation. Most businesses have reached the peak of efficiency and effectiveness just as the need for innovation becomes far more important, and many of the prized investments in efficiency and effectiveness present barriers that make the necessary shift to innovation more difficult.

In hindsight, we've arrived at a fairly logical end state. From hierarchical management structures to Taylor's Scientific Management to TQM, BPR, and on to Six Sigma and Lean, the vast majority of tools and techniques and the best management thinking have been focused on productivity and efficiency. Our firms are models of efficiency, sources of wonder for many economists who cannot understand how firms still achieve even more productivity from overworked but highly efficient business models.

Yet at the height of this achievement, businesses face an important shift in strategy. The historic management focus and investment has left many organizations locked into rigid operating models at a time when quickness and flexibility are increasingly in demand. New competitors in low-cost markets, heightened competition, and an increasingly demanding consumer base make it ever more difficult to squeeze out more savings. At the same time, traditional low-cost competitors are rapidly gaining in product development, innovation, and creativity capabilities, threatening to eliminate all sources of competitive advantage for many U.S.-based firms. Let's look at why some firms, the relentless innovators, can innovate successfully, consistently over time, and what your firm can learn from the industry leaders.

Chapter 4

What Relentless Innovators Have in Common

In any human endeavor, participants are sorted out according to abilities. Whenever we examine human exploits such as business or education, there is always an inevitable hierarchy that occurs. Achievement in athletic events provides excellent examples of this fact. Let's consider basketball.

Basketball is played around the globe, with the height of achievement to join the United States National Basketball Association (NBA). Almost 350 colleges in the United States participate in Division 1 basketball.[1] Each team has, on average, 12 players. Of these approximately 4,000 athletes, many hope to play in the NBA. Beyond Division 1 there are hundreds of other colleges and universities that field teams; plenty of NBA players hail from the ranks of these schools. Further, thousands of teenagers and young adults play basketball in colleges and in clubs in Europe, South America, and Asia, with similar dreams of reaching the ranks of the NBA.

Yet in any year, 60 players get drafted,[2] and of those perhaps only half actually make the professional rosters. Of those who make it into the pros, only two or three players on any team claim the lion's share of professional compensation.[3] As in all professional sports, most of the rewards accrue to a small number of the actual participants.

The same sorting is seen in other fields as well. In academics, for example, just over 3 million students enter college each year as freshmen in the United States,[4] and of those less than 40 percent graduate with a degree four years later.[5] An even smaller percentage of those with undergraduate degrees go on to graduate school. In case after case, it's evident that in any endeavor, a small handful of individuals, firms, or teams accrue most of the recognition, rewards, and benefits. The same is true in business.

The Standard & Poor's 500 Index was established in 1957 to track leading publicly traded firms. While the United States has more than 15,000 publicly traded companies, only 500 are closely tracked, and of these only a handful are consistently profitable over time. Of the 500 firms in the original S&P 500, only 125 still existed in 2003, according to an article in Investopedia.[6] Of the 125 firms still in existence, only 94 of those firms remain in the current S&P 500. In just over 50 years, 375 of the original 500 firms were acquired or went out of business. To understand how to maintain success and flourish over the longer term, it is best to investigate firms that have managed to do so already.

Innovation Ranking

The same natural ranking found in sports and academia can be applied to innovation. Thousands of firms in every region of the globe strive to create interesting, valuable, and original products and services. Many achieve that goal once, only to be acquired or to fail at developing a second product or service. Some grow and scale, achieving financial success, but never become a leader in their industry. Other firms find startling successes periodically, creating interesting new products, but they cannot sustain the creativity necessary to remain innovative over time.

Only a handful of firms, the ones constantly referred to in the case studies here, the relentless innovators, have sustained innovation leadership over decades. Apple, Google, 3M, W. L. Gore, and Procter & Gamble can be thought of as sustaining innovation leadership over a long period of time, in varying market conditions and geographies, with a range of product offerings, and with different leaders. Consider these questions while reading this chapter, and those that follow:

- What makes these firms successful over the long run, while most firms innovate rarely, if at all?
- Is there a cap on the amount of innovative ideas or people in an industry or geographical area?
- Do innovative people naturally gravitate to a few specific innovative companies?
- Is innovation success based on visionary leadership, excellent R&D skills, leading technology?

- Does innovation only happen in industries with heightened competition, or only in entrepreneurial firms that seek to disrupt existing markets?
- Is there a natural limit to the number of innovative firms that can exist in an industry, or are there other factors at play?

Apple, Google, and the other firms named have been studied by a range of expert innovators and academic thinkers. For example, there have been a number of rationales put forth as to why Apple innovates consistently, while Dell hasn't produced a large-scale innovation since it defined a new business model for the personal computer.[7] In fact, when asked in 1997 what he would do if he were the CEO of Apple, Michael Dell said "What would I do? I'd shut it down and give the money back to the shareholders." Today Apple has a market valuation 10 times larger than Dell.[8]

Studies have also looked at why Procter & Gamble consistently creates valuable new products in a range of markets, geographies, and cultures, while other consumer goods companies create far fewer. Procter & Gamble, as an example, leverages "open" innovation with its customers and partners more effectively than many of its competitors. Open innovation is a relatively new application of innovation, which places emphasis on obtaining ideas and technologies from customers and business partners, rather than generating all of the ideas internally. Open innovation was first described by Henry Chesbrough in his book *Open Innovation*,[9] and in the last decade it has grown as a viable innovation technique. There are several forms of open innovation, including "crowdsourcing" and proprietary networks. Crowdsourcing provides a means for your customers and partners to suggest ideas to your firm—almost a virtual "suggestion box" over the Internet. Procter & Gamble leverages the concept of proprietary networks—trusted business partners that can provide new ideas or new technologies to create new products and services.[10]

Is the adoption of a specific innovation method or technique the driver for *sustained innovation*? Simply put: no. In fact, there are forces far stronger than specific innovation tools or methodologies, factors within your organization that can be harnessed to create more innovation on a consistent basis; these skills and people already exist.

Unlike the basketball and educational analogies used earlier, however, I don't believe the number of innovators is capped. In fact, looking at the handful of relentless innovators identified above, they share almost nothing in common, other than their ability to innovate consistently over time. Further, since stripping away much of the mythology and exposing the truth that only your BAU model and middle managers impede the adoption of more innovation, any firm of any size in any industry can become more innovative with enough focus and determination.

There are several reasons that firms don't innovate successfully over time that I examine here, many of which are symptoms of one of the two main culprits already discussed: business as usual.

Why Firms Can't Sustain Innovation

There are at least four key reasons that innovation isn't sustainable in an organization related to the "business as usual" barrier. While I assert that the BAU barrier is *most* important, it is instructive to review and address these factors as well.

Strategic Vision and Communication

Often, innovation is thought of as a "strategy," rather than an *enabler* to strategic goals and objectives. Innovation can certainly improve strategic outcomes, but only if those goals and objectives are well defined, understood, and communicated throughout a firm. Executive teams typically form strategies that are either poorly communicated to the organization, or are well communicated but aren't especially clear. Every firm creates strategic plans, but they face two significant challenges.

First, the plans must define what the firm will do, and what it won't do. As a firm grows and adds products and services to its portfolio, executives are often challenged more by defining what the firm *won't* do than by what it *will* do. Strange as it may seem, large firms have great difficulty focusing on the markets and customers in which they can be most successful, and often they expand into markets, geographies, and industries in which they can't compete effectively. This expansionist mentality exists because it is often easier to enter new markets than to think carefully about growth and innovation in specifically targeted markets or segments. Only a clear strategy that defines specific markets, industries, and geographies keeps an organization from entering industries or customer segments not strategic to the firm.

Second, the strategic plans must be well communicated and distributed internally in order to be deployed. Strategic plans are rarely explained effectively, which leaves them open to interpretation. Since innovation introduces new tools, goals, and risks, clarity about the strategic direction of the firm is paramount to ensure that innovative efforts don't pursue interesting but *irrelevant* technologies, products, and services. In the instances in which innovators lack clear direction about corporate strategy and goals, they frequently pursue ideas that seem valuable and appealing, only to create products and services that don't align with the poorly defined or poorly communicated strategy. This outcome is a main reason that executives are uncomfortable with innovation: the frequency of "disconnects" created between what executives want and what innovators produce. This problem is laid at the feet of the executives, who are responsible not only for creating a viable strategy but also for communicating those goals and strategies clearly throughout the organization.

Strategic clarity also suffers when firms are so generously rewarded for *operational consistency*. The financial markets reward companies that regularly crank out predictable returns. Therefore, in many cases, interesting, expansive strategy gives way to one strategy designed to sustain consistency. One can easily argue that the core strategy of most firms is to maintain consistency of operations rather than to introduce new products or services. As discussed, this focus on consistency, in the absence of any divergent strategies clearly communicated and reinforced by executives, has the unintended result of strengthening BAU. Innovation within this context is almost impossible, since there are no clear targets, no stomach for divergent thinking, and the investment in the existing operating model overwhelms any small innovation sparks that may occur. Valuable innovation cannot happen in the absence of clear strategy.

> Valuable innovation cannot happen in the absence of clear strategy, well communicated and constantly reinforced.

An innovation "failure" in this instance is due to poor strategic definition and the communication of those strategies, rather than an inability to innovate. This common innovation barrier, however, can easily be overcome by creating a "charter" that innovators validate with the sponsoring executive.

A charter is simply a statement of strategy, scope, and intent, and it is agreed upon by sponsoring executives and the individuals who will complete the project. In the absence of clear strategy and scope, the innovators themselves may need to develop the charter and have it approved by an executive team. The charter should consist of several key components:

- A clearly defined problem to solve or opportunity to address
- The expected scope of the innovation effort, that is, how incremental or disruptive the ideas should be

- The executive or team responsible for sponsoring the work
- Timeframes and milestones associated with the effort
- Some definition of the end product or result

Once developed and approved by an executive or steering committee, the charter serves as a basis for defining the scope of the work, and as an easy reference to ensure that the ideas generated meet the goals of the original project.

Remember, the goal of the charter isn't to dictate the specific outcome, but to define the intent, goals, scope, and work necessary to solve an important problem or address an emerging opportunity. The charter helps reassure executives that the results will be in line with corporate goals, and it dictates a helpful scope for the innovators at the same time.

Inadequate Resourcing

The second reason many firms can't innovate successfully over a sustained period of time is inadequate resourcing. Since innovation is often poorly understood, it is also poorly supported, both from a financial perspective and a headcount perspective. Far too many innovation projects have little to no budget, since they are not given priority. In addition to this lack of funds, innovation teams have inadequate staffing, causing them to "borrow" people who have critical knowledge from other teams or departments.

Innovation roles are rarely full-time roles. Innovation leadership is often assigned to managers in addition to their regular jobs, which means the task is merely tacked on to an already hectic schedule. Since few firms have much innovation experience, managers and executives underestimate the work and resources required to innovate successfully. This lack of understanding leads executives to appoint "part-time" managers who must further beg or borrow resources.

Compounding this error, many innovation projects have unrealistically short timeframes because managers estimate innovation efforts based on known project timelines. These plans don't incorporate the learning curve associated with innovation, the experimentation that's necessary, and the barriers that the team will encounter doing something completely new. Given these constraints, many savvy managers believe the effort is doomed from the start and avoid innovation at all costs.

When the effort and magnitude of the innovation effort is understood, however, it is possible to obtain more resources and funding to complete the initiative successfully, or change the expectations or scope of the project. Inadequate resourcing will doom any project. Since innovation activities are so poorly understood, they are often underestimated and under-resourced, which is another significant reason that so many innovation projects "fail."

Accepting a poorly defined and scoped innovation effort with inadequate resources is a recipe for abject failure, so take the time up front to educate your management team about the work involved and the resources required. The lack of comprehension about the work involved in an innovation effort, and the subsequent lack of resources, will derail an innovation effort more quickly than almost any other factor.

A "Project" Rather than "Capability" Mindset

A third reason many firms can't innovate successfully over time is because they think of innovation as a discrete "project," intended to build a new product or service, rather than a capability that can be developed, repeated, and reused. If the intent is simply to respond to an external threat or opportunity, and not to innovate continuously, there's little rationale for building a robust innovation discipline or capability.

The firm that innovates once but doesn't document its processes and methods loses the hard-won innovation expertise and knowledge quickly as the people who were involved in the efforts return to their regular jobs. Contrast that approach with firms where people are trained in specific roles and benefit from successive innovation projects in which a learning curve is established. This consistent innovation provides the ability for teams and projects to leverage prior experience over time. In a firm that considers innovation a discrete project to be accomplished once, those skills and knowledge are never developed, so every innovation effort is customized from "scratch" and unable to be sustained.

In any task or process, people gain skills and competence by following a consistent process over time. If those involved don't follow a process and they can't retain the expertise and learning, every innovation effort is a completely de novo experience. As new teams with little prior experience and few milestones or roadmaps struggle to innovate, they are often labeled as "failures." This failure isn't one of innovation, though; it is a failure to *capture* and *reuse knowledge*. The effort associated with defining a consistent innovation capability pales in comparison to the effort to start several distinct projects. The skills and capabilities gained by following an innovation discipline will accelerate all of the subsequent innovation projects.

Defining and building an innovation process is expensive, but doing so provides a foundation on which every other innovation project can rest. Otherwise, every initiative must design and build its own foundation, which wastes time and resources. For firms focused on efficiency and effectiveness, innovation as a consistent business discipline should be a first priority; otherwise every innovation effort results in inefficiency and rework.

Defining an innovation process requires identifying the important steps in an innovation activity, usually:

- Spotting trends to understand potential future needs and opportunities
- Gathering customer insights, needs, and jobs to be done
- Generating ideas based on these inputs
- Managing, developing, and evaluating ideas
- Prototyping and piloting ideas
- Selecting the best ideas to commercialize
- Converting ideas into new products and services
- Launching new products and services

Each phase consists of a number of activities, and those activities require specific skills and roles. Defining a consistent innovation methodology and the inputs, outputs, and roles necessary to complete the activity will simplify innovation and allow your organization to increase its skills and innovation competence. See the recommended readings at the end of this book for several books that provide more information on designing and building an innovation process.[11]

Control and Risk

A firm that hews closely to its "core competence" and has an established set of products and services is easier to manage and understand than a firm that has many innovation initiatives. As the number of initiatives and products grow, executives have greater difficulty understanding the business and communicating its value proposition to external markets; they are then far more likely to "surprise" the market with results that don't achieve expectations. Markets react with great speed to downgrade the stock of a firm that fails to achieve expectations, so executives are careful never to surprise the market. It is far easier and far safer to manage a business that sticks to its core competencies, sustaining an established line of products or services, than it is to lead a firm deeply invested in innovation. Financial markets hate uncertainty and surprises, and they have taught executives to avoid them. Since innovation introduces these elements, it is even viewed with suspicion by the executives who demand it! Many firms fail to innovate because of the reluctance to surprise the market or the failure to meet market expectations. This "failure" is a failure of management's understanding, control, and bandwidth, not a failure of innovation tools, techniques, or methods.

For these reasons and many more, innovation is often treated as an "exploratory" effort, assigned to a small team that is understaffed, isolated from the rest of the organization with unclear goals. Rather than view innovation as an important initiative that can drive the success of the company, many innovation efforts are half-hearted efforts that don't engage the businesses effectively and don't include the best people in the business.

Let's examine these issues with the following example.

A Company's Struggles: Part 1

Think about a company whose focus for many years has been on delivering consistent results. Let's say it is a Fortune 500 firm, with operations in North America, Europe, and Asia. Its headquarters are in Chicago and it has three lines of business: aerospace, machined tools, and pumps. These products are designed in the United States and Europe, manufactured in the United States, Europe, and Asia, and distributed worldwide. The company has an individual responsible for North America, Europe, and Asia, and a person who heads up each of the three lines of business.

Due to slow economic growth, the firm has seen little revenue growth year on year, so the quarterly results have been achieved through radical cost reduction and efficiency and productivity gains. The CEO knows he needs to spark revenue and profit growth, but he isn't sure how to direct the teams to create ideas that may lead to fresh products and original services that will drive additional revenue. Lately he has begun to meet with his executive team and ask "How can we become more innovative?"

The individuals who head up the product teams would be happy to create new products, but they don't have the staff or resources necessary to do so consistently, and, frankly, such creativity has fallen out of practice within the groups. They lack insight about existing customers and they don't have good information about potential emerging customer segments. Further, the existing business model stands in the way of interesting and valuable innovation because it relies on a big initial sale of capital equipment and high margin maintenance services. Many of the new entrants have developed different business models that have attracted new customers, yet the company's predictable quarterly returns are predicated on the existing business model. Changing the business model could negatively affect the consistent achievement of financial goals and play havoc with the stock price.

Finally, some leaders in the product lines are concerned that innovative new products and services may disrupt existing ones, perhaps reducing the executives' influence and power base. A certain amount of defensiveness has crept in, as executives who lead product groups and geographic lines of business react to the increasing pace of change by developing a defensive posture, meant to deflect internal and external threats, rather than to identify the threats and seek to adapt to them. Like many of its peers, this company has consistently "doubled down" on efficiency and effectiveness, resulting in a finely tuned operating model capable of generating consistent quarterly results, but leaving the organization short of ideas and of the capability to turn ideas into new products and services. The CEO realizes this situation and he is perhaps most frustrated by the fact that no one in the organization seems interested in spearheading a new innovation effort.

Why No One Wants to Lead Innovation Efforts

The reasons discussed so far all relate to the business-as-usual operating model. These barriers cause innovation often to be treated as an "exploratory" effort, assigned to a small understaffed team, isolated from the rest of the organization, with unclear goals. Rather than view innovation as an important initiative that can drive the success of the company, many efforts are half-hearted, they don't engage the businesses effectively, and they don't include the best people the companies have to offer. Innovation remains the most people-intensive activity in a firm, and the "best" employees are vital for success, yet there are several reasons that strong people avoid innovation projects. In fact, most innovation projects are likely to be dead ends for high-potential leaders. Even though senior executives regularly expound on the need for innovation, it can be difficult to find a viable executive or senior manager to lead an innovation effort. The reasons won't surprise you.

> Many of the best managers don't want to run an innovation effort, primarily due to the risk of failure, the lack of resources, and uncertain support from executives.

Inherent Risk of Failure

In an innovation setting, "failure" is often a learning experience, creating useful insights for future efforts. In many corporate settings, however, "failure" in any form is a career-limiting outcome. No one climbing the corporate ladder wants to be associated with initiatives that have not worked. Most executives and managers understand that innovation is inherently risky and subject to failure, not only because of its uncertainty, but also because the strategies aren't fully formed or the necessary resources allocated.

In a corporate environment in which any failure is looked at as a blemish on one's career, no executive on the fast track is likely to seek out an innovation leadership role. An executive tasked with building an innovation team will demand the most experienced team he or she can possibly acquire, and he or she will be faced with one of two outcomes: an apprehensive team full of experienced people who were assigned to the effort but don't have their hearts in it, or a relatively junior team of volunteers who believe in innovation but don't have the optimal experiences or connections. The best innovation team, however, combines passion about the opportunity and deep experience within the firm. Without these two elements, failure is imminent.

Unfamiliar Tools and Techniques

Running an innovation project also means using unfamiliar tools and techniques. During the downturn and recession after 2008, many industries trimmed training budgets as a way to cut costs. Even when training budgets were larger, there was little training available on creativity and innovation. In many innovation projects, little training is provided for the innovation team, but the team is still expected to generate interesting, relevant, disruptive ideas in a relatively short period of time. This lack of training compounds the fact that apprehensive or poorly equipped people are forced to use unfamiliar tools and techniques to create risky new propositions, while magnifying poor leadership and unwilling participation.

Team Size

Most innovation project teams are small, due to some of the constraints we've identified above. However, an executive's power and status are often signaled by the number of people, or the size of the budget, he or she manages. Typically, the higher an individual climbs in an organization, the more people, dollars, and products he or she is in charge of, which is sometimes referred to as their "pyramid." Seniority and expertise is reflected in the size and shape of this pyramid.

For many of the best, most experienced managers, leading an innovation project can be viewed as a significant step down from their current posts, where they manage more people and larger budgets. Many executives are therefore reluctant to lead an innovation project. This reluctance is unfortunate but understandable given compensation, evaluation, and career track considerations. Most managers and executives can't afford to step away from their core competencies, their career track, and their ongoing projects and initiatives. There are too few management "slots" at any level. Leaving a role that manages a large budget or large team to run a short-term smaller project can leave a manager or executive without a role once the innovation effort is complete.

The result is that many innovation projects are led by junior managers or executives who have less power, less experience, and fewer connections in an organization than is optimal. While innovation initiatives may offer a challenging "growth" opportunity for a junior executive, the lack of experienced leadership simply adds to the risk and uncertainty of an innovation project. Additionally, many of the executives or managers who could lead an innovation effort have a number of projects underway that they have developed and that they want to see succeed. They believe that shifting their focus to an innovation effort will divert attention from key projects and initiatives they've nurtured over time. The executive rightly fears that these carefully nurtured projects may be diverted or fail due to a lack of adequate management.

> Experienced managers and executives don't want to run innovation efforts because the role can be viewed as a step down from the pyramids or budgets they usually manage.

A Company's Struggles: Part 2

Let's look back in on our company discussed earlier. The CEO has had informal discussions with several senior executives who are direct reports, seeking their advice and input on forming an innovation team. Each product executive has expressed interest mingled with concern. None of them want to kick off an innovation effort unless the CEO will help fund and resource the effort, and all of them are concerned about finding the right people to lead the team.

Furthermore, none of the product executives from the three business lines want to take their best people off their current tasks, since that would likely mean the "core" work wouldn't be completed effectively and important goals or customer commitments would be missed. Each executive knows it will be hard to recruit a senior individual to an uncertain, risky task aimed at developing new products and services. Few capable leaders exist, and most of them have more work than they can handle now. Plus, to take on an innovation role would be a significant step down in terms of the number of people they manage and the budgets they are responsible for.

After taking the pulse of their direct reports, the heads of two business units, machined parts and pumps, can't identify a candidate to lead an innovation effort they believe is experienced enough and has the respect of the organization. The aerospace division has identified a person to lead an innovation effort, a fast-rising young executive, but it is clear that he will need help and guidance as his support base and connections aren't sufficient to get everything done he'll need to do. However, since he didn't have a large team he wasn't averse to leading the innovation effort, and he sees the opportunity as a springboard to advancement if he is successful.

Any Firm Can Build an Innovation Discipline

Is there a limit or cap to the number of firms that can achieve relentless innovation? Are firms like Apple, Google, 3M, and W. L. Gore so unique in their leadership, culture, strategy, or other attributes as to stand alone as relentless innovators? Do these firms have better management, better insight, or smarter employees than other firms?

The answer is no.

These firms have little in common with each other, other than the ability to innovate consistently. Some of these firms have well known, dominant leaders, and others have no clear leader at all. The organizations compete across varying industries and they are involved in different technologies and markets. Some are old, established businesses in competitive industries, while others are new, rapidly growing companies in industries that didn't exist just a decade ago. Why can Google innovate and why does your firm struggle? Why does Apple seem to be able to create innovations almost on demand, while your firm doesn't innovate well even under pressure? Do corporate structure and organizational memory play the biggest role? Just because Apple's and Google's founders are still involved in senior roles, do they have one up on your company?

While all of these factors are important, I believe they aren't nearly as important as this: what these firms all share *is an operating model tuned to innovation.* This operating model can be adopted and learned by any other business that seeks to become a relentless innovator. The concepts of an innovative operating model, which is one component in the success of these firms, have been addressed. Now, let's return to the observation and discussion of those people who develop, support, and enable a viable operating model: middle management.

Chapter 5

The Antagonist

Every story has a protagonist, the leading man or woman who is the "star" of the show. Considering the "story" of innovation in corporate settings, let's consider the "middle manager" as the protagonist. I've argued that middle managers, though unloved and underappreciated, hold the most important role in any organization. Middle managers work heroically to achieve profit, driving results and using inputs efficiently to sustain high share prices with fewer and fewer resources, yet their recognition and rewards pale next to those of executives. They strive to do what's best for the company in terms of effectiveness while delivering a consistent quarterly result. These individuals are working to achieve the goals as established by the executive team while keeping the organization humming. They are beset on all sides—by unhappy customers, disruptive partners, market disruptions, disgruntled staff and, perhaps most important, by executives with new needs and requirements.

In classic literature, as well as in life, protagonists have a corresponding antagonist, a person or problem that forces the protagonist to greater achievements. This is the classic setting for a story, and one that is true in an innovation setting now as much as it was true in the Greek amphitheater thousands of years ago. The *Merriam-Webster Dictionary* defines an antagonist as "one that contends with or opposes another." A senior executive is the best example of an antagonist in regard to middle management, especially when that executive needs innovation to create new products or value for the business.

Uncomfortable Executives

Most executives would prefer to take other avenues to growth and differentiation than innovation, if the truth is told, because innovation makes many of them uncomfortable. It took years of experience for me to arrive at this insight.

> While executives regularly expound on the need for innovation, in reality, innovation makes many executives uncomfortable.

Many executives are concerned about the impact innovation will have on their organization. They recognize the potential benefits innovation can deliver but the difficult and sometimes elusive path to these benefits leaves them uncomfortable. As I've worked through this revelation, I've documented at least three reasons that innovation makes executives feel this way.

Uncertainty

First, innovation is uncertain since it will force the firm to pursue new products or business models that may not deliver the desired results and that may distract the organization from its core business. Just like middle managers, executives realize the value of consistent performance and efficiency. Anything that introduces uncertainty into the operating model also introduces inefficiency and doubt. Executives have become comfortable with management tools like Six Sigma, reporting quantitative results to several decimal points. Innovation, however, is often qualitative and it can't be adequately reduced to pure numbers or facts. It requires better insight into unmet or unspoken needs than that of competitors, or even customers. Often, there's simply no way to validate an idea to the degree that executives have become accustomed to in other initiatives. So, innovators must launch out on faith based on their instincts about their ideas, or go through a time-consuming process to gather more information for validation, which may lead to a loss of differentiation and leadership. The lack of quantifiable information means innovation is risky and uncertain. Ideas and even products are subject to failure, and no executive wants a major initiative to fail on his or her watch.

Innovation also introduces uncertainty because it requires new tools and techniques that are unfamiliar. As I've noted before, over the last 20 years, the operating models have been honed to ultimate efficiency. This efficiency is based in large part on well-understood tools and techniques. These familiar tools don't introduce any uncertainty in daily operations, but innovation does. Any activity that introduces uncertainty introduces risk, inefficiency, and distraction, so executives are wary of innovation initiatives for this reason.

Financial Resources

Beyond uncertainty, innovation makes executives uncomfortable because it doesn't fit in the planning and budgeting scheme. Most organizations have standard, familiar investment buckets. Annual plans look similar year on year. Money for innovation, however, hasn't been allocated in previous budgets and has no "bucket" in the budgeting process, so it's rare that funds are set aside for such initiatives. Since there's no historic basis for innovation funding or budgeting, it is difficult to determine whether or not innovation should be funded, and, if so, who should receive funding and in what amounts.

Executives aren't naïve. They realize the benefits innovation can provide, but they also acknowledge that those benefits can only be recognized in the distant future, while the costs associated with innovation projects happen in the current year. Innovation efforts have an immediate, negative impact to the bottom line long before they have the potential to deliver a positive result in terms of new revenues and profits. It is difficult for any executive to agree to fund initiatives with long-term payoffs and short-term costs, especially an initiative that introduces risk and an uncertain payoff.

Human Resources

Further, beyond financial resources is the question of human resources. After years of outsourcing and downsizing, most organizations have little management slack and even fewer have executives or managers with innovation experience. In organizations in which the operating model is honed to high efficiency and effectiveness, it is difficult to shift any human resources to other efforts, since many firms operate on the bleeding edge of efficiency. Assigning employees to an uncertain and risky innovation effort means reducing staffing and focus for product lines or operating units that have already demonstrated excellent performance. Although the outcomes may be uncertain, real resources must be invested to achieve any innovation outcome. Many executives prefer the safety and certainty of investing the workforce in known products or services.

Predicting the Outcome

Since innovation is risky, new, and uncertain, the results of an innovation effort can be difficult to predict. In an era when predictability is the hallmark of an excellent executive, at least in the financial markets, innovation's unpredictability leaves executives exposed to perhaps the worst outcome of all—surprising the financial markets with unexpected financial returns. Since the compensation of many executives is tightly tied to their stock price, and Wall Street values firms that produce consistent quarter-on-quarter results, surprises, *even positive ones that drive new earnings*, aren't usually welcome. The more innovation underway in an organization, the less predictable the results become. More unpredictability leads to more work for the executive team to understand the potential outcomes and communicate them to external parties, especially shareholders. In many instances, innovation can create more work and headaches than benefits for executives.

How a Senior Executive Becomes an Antagonist

For most executives, innovation is a last resort when their products or services are under attack and they have considered and deployed all the other tools at their disposal. But regardless of their feelings or beliefs about innovation, executives will embrace innovation once all the other possibilities have been exhausted; they have to. Whether they come to believe in innovation based on what they read about the success of other firms, or what they witness in the markets, or they simply turn to innovation out of desperation, a few executives in any organization will decide that innovation is simply too important to ignore. And when they make that decision, they become a potential disrupter to the business-as-usual operating model and therefore an antagonist to the middle manager.

Executives have some understanding of the challenges innovation presents to their business. They know that the organization is a well-oiled machine, and optimized to do what it does correctly. They understand that there are few people with much innovation experience, but they also recognize that many firms, in their industry and others, have successfully created new products and services while driving new revenues and profits by working to become more innovative. So, in spite of the acknowledged resistance to innovation and the lack of knowledge and capability in their business, they decide the potential results of innovation are more important than the barriers that exist within the organization. This decision places their goals at odds with the middle managers, who are supposed to protect and sustain the BAU processes.

Executive Knowledge

Executives, for their part, have only a rudimentary knowledge of how work actually gets done in their organization. Yes, many of them worked in the lower levels and rose up through the ranks, so they have experience in the operations, but the emphasis over the last 20 years has been on efficiency, not innovation. In fact, over the last two decades, it is far more likely that these executives demonstrated their capabilities as cost cutters, rightsizers, and efficiency experts. From their time on the "front lines" they may recall their organizations as being run fairly efficiently, but it is likely that those firms also had additional resources and capabilities that no longer exist today.

Many executives don't appreciate the continual refinement of the operational processes. They also may not know the extent to which many processes are functioning on the bleeding edge of efficiency, with no fat left to cut. This efficiency means the firm operates at the lowest possible cost, but there is simply no "slack" in the system to take up new projects or initiatives. Most middle managers and staff are booked far beyond their traditional 40-hour work week simply *sustaining* their regular jobs. Attempting to add another initiative, especially as disruptive as an innovation effort, is exceptionally arduous—few executives realize the distractions and difficulties involved in their request.

Further, since most executives haven't had much experience innovating as they climbed the corporate ladder, they don't appreciate how little internal innovation skill their organizations possess, how little training has been deployed, and the lack of emphasis on creativity and innovation. The significant focus on corporate innovation as a consistent discipline is a relatively new phenomenon. Cost-cutting and right-sizing has been reinforced and rewarded; little training or skill development has been spent teaching people about creativity and innovation. The overwhelming focus on efficiency and the lack of focus on developing innovation skills means that most organizations have very little internal capabilities or knowledge about innovation, its methods, and its techniques.

Even if those called on to conduct an innovation initiative had the time available in their schedules, they likely don't have the necessary experience or familiarity. While it's evident to readers as impartial observers that middle managers have become "one dimensional," many executives fail to understand how little knowledge, skill, and passion there is for innovation in their MM ranks. Since middle managers won't have time to learn innovation techniques, they'll fall back on the techniques they already know, simply reinforcing BAU and attempting to force innovation initiatives into a process that doesn't accept new ideas.

Lack of Definition

Executives often doom innovators by asking for innovation without creating clear definitions or the expected outcomes of the effort. For example, does the firm need "disruptive" innovation—radically new products and services? Does the executive want to implement "open innovation" as a technique to work with partners and customers to gather and develop ideas? Does the executive have specific goals?

As an example of a defined innovation initiative, 3M has a stated goal of generating 30 percent of its revenues from products launched in the previous four years.[1] This straightforward, constantly reinforced objective is an example of executive management sending clear signals about the importance of innovation. Such clarity from a firm, however, is fairly unusual. The lack of clear communication about innovation leaves middle managers and innovation teams in a tremendous bind. Since innovation is an uncertain task, taken on by managers and staff who lack necessary skills and expertise, defining a clear objective and scope helps shape the work and provides focus. A poorly defined objective or scope leaves an innovation team spinning their wheels for weeks just trying to agree on what to do, recognizing that they don't have the time, resources, or knowledge to innovate even if they do come to a conclusion on what the executive wants.

Lack of Follow Through

Another reason executives are often antagonists in an innovation story is their propensity to stake out a vision, demand innovation, and subsequently spend little time or effort to ensure the work is carried out. Executives in the firms we've profiled as relentless innovators actively involve themselves in the innovation work by

- Spotting new needs and defining new products (Steve Jobs at Apple), or
- Encouraging innovation teams to develop new products (Terri Kelly, CEO of Gore), or
- Reinforcing the importance of innovation with executive leadership and in the corporate culture (Buckley at 3M).

Yes, it is a hackneyed phrase but one that is simply too important to ignore: executives must "walk the talk" where innovation is concerned. They must be far more engaged in the effort, becoming an innovation champion, cheerleader, and financier, in order for any long-term, sustainable initiative to take root. Simply demanding that the firm "become more innovative" without investing time or personal capital only confuses the organization and creates cynicism about the request.

The Results of Innovation without a Plan

Certainly you can begin to see why senior executives who demand innovation from their product groups or other teams are antagonists to middle managers and the BAU process. While executives may need innovation to achieve growth goals, they introduce a major threat to business as usual *without* a corresponding increase in staffing, resources, knowledge, or funding.

Middle managers may simply ignore the executive's innovation request, hoping that the proposed effort is a knee jerk reaction to a competitor or market condition that the executive will quickly drop. Executives are pressed for time and typically assume their requests will be acted on. Weeks or even months later as they return to the issue, executives are surprised and disappointed that little was accomplished. In these cases, however, middle managers are probably correct in deflecting the request until it is demonstrated that the need is real.

As noted previously, many executives simply don't understand the resources and investment necessary to complete an innovation activity, so middle managers' requests for funding, staffing, and resources may seem unreasonable. If executives and middle managers actually discussed and debated the needs associated with the request, executives would learn more about the resource needs and, subsequently, innovation could occur. More often than not, though, executives believe their teams are unwilling to innovate, so they drop the request or turn to outside resources for assistance. This situation can become exacerbated when middle managers accept the innovation request but only provide as little commitment as possible.

Executives want quick "wins" with little cost or distraction and they don't plan to change the processes or culture of the business. This approach leads to great cynicism, as innovation becomes a discrete event that disrupts business as usual rather than an internal capability. Executives simply want to discover a new product or service that customers need, and offer that product or service quickly, with as little investment, time, and distraction as possible. After all, executives rarely budget for innovation, so there are typically no funds or resources set aside for these efforts. Also, every executive understands the effort involved in changing corporate "culture," so no one executive is going to advocate change of that magnitude or level. No executive can afford to distract his or her organization with a temporary focus on innovation if that focus distracts the firm and results in worse than expected quarterly results.

Executives hedge their bets, seeking one or two great ideas from their staff and middle management, without funding, resources, or training. In their heart of hearts, executives know the challenges they are presenting to middle managers and the status quo, but executives don't have the time, the patience, or the desire to make sweeping changes. Creating a few good ideas with minimum impact to the organization and with as little investment as possible is what's desired.

If creating truly innovative and disruptive ideas were easy, those ideas would be manifest in the organization as new products and services. Everyone understands from the beginning how difficult it is to create compelling new ideas in any situation, much less to convert those ideas into viable products and services. To compound the difficulty, executives are asking for disruptive ideas while expecting the business to continue to operate at full effectiveness and efficiency. Middle managers receive these messages and understand the unspoken dichotomy in the request: create radical, valuable new products and services but don't upset the status quo. Everyone involved in this transaction understands how difficult it is simply to achieve quarterly goals within the business-as-usual framework, much less attempt to create valuable new ideas. The best that middle managers can do is to create the appearance of an innovation initiative or project. The consistent result is that no one is satisfied, no good results are created, and "innovation" takes the blame.

Who's to Blame?

When executives request disruptive ideas but they don't define the desired outcome and fail to offer appropriate tools or resources, they play the part of the emperor in the fable of the emperor's new clothes. Everyone recognizes the dissonance of the request, but few have the temerity to alert the emperor, or in this case the executive, to the nature of his or her request.

Certainly the executives can't point the finger of blame at themselves, because that would admit that the request was poorly conceived, or inadequately funded, or that the executive simply didn't understand the magnitude of the request. They can't, however, reasonably point the blame at middle management either because while the ideas may not have been interesting or valuable, the middle managers delivered on their most important tasks—delivering the anticipated quarterly results.

The only whipping boys left are the skills and capabilities of the "people"—"our people aren't very innovative"—or the inability of innovation tools and techniques to deliver results—"innovation never delivers." Executives must either stipulate that their organization tried its best but simply couldn't innovate effectively because of the internal capabilities of the organization or they must blame innovation tools and techniques as charlatans that never deliver value.

> When innovation "fails" executives are left with two responses: "Our people aren't very innovative" or "Innovation as a technique doesn't deliver." Neither is true.

As executives take the role of the antagonist, no one is quite sure who to blame for the lack of innovation success. Great cynicism about innovation and about the executives who request innovation but who don't support it or fund it is instilled in middle management. In the meantime, executives don't know what went wrong, but they may also not care as long as the numbers are met, leaving innovation to fall by the wayside. This cynicism and lack of support, combined with the real need for more innovation, create the crisis that we'll examine in the next chapter.

Relentless Innovator Executives

Contrast the portrait I've painted of the typical executive with executives in companies that are relentless innovators. While I've argued that executive leadership is not the driver for sustained innovation, in firms that follow an innovation BAU operating model, executives recognize the importance and value of innovation and encourage the organization to build skills, methods, and infrastructure to enable everyone to innovate repeatedly.

Executives in companies like P&G or Google are constantly reinforcing the importance of innovation, establishing it as part of the organizational belief system or culture. These executives *actively participate in innovation activities*, investing their own personal capital in the innovation effort and demonstrating how valuable they believe consistent innovation is to their businesses. Rather than an "us versus them" mentality, executives work *with* middle managers to define innovation objectives and ensure that the skills, knowledge, and capabilities exist to develop an innovation discipline. In this case, the alignment between what executives want and expect, and what they are willing to invest, is balanced. Look no further than 3M and the replacement of McNerney by Buckley. Within a short time of his arrival, Buckley was shifting the operating model back into balance, bringing more emphasis and resources to bear on innovation. Executives in these relentless innovators understand the power of clear strategy and communication. Lafley's commitment to open innovation meant that Procter & Gamble had to shift resources, processes, and skills to embrace open innovation. Executives in relentless innovators encourage an innovative environment. They create cultures that embrace innovation and risk taking at all levels of the organization. Both Google and 3M practice 15 percent time, giving employees time to dream up new ideas. Gore encourages its employees to create new products and services based on its key technologies. These firms and their executives demonstrate that their innovation goals will be supported over the long term—innovation isn't simply a one-time need or event, but part of the fabric of the way they lead.

Steve Jobs at Apple presents the most compelling example of an executive who understands the need for innovation, and the internal commitment and focus required to create interesting new products while maintaining high efficiency. Jobs and his executive team are active in every new innovative product or service. One of Jobs's first acts as the new head of Apple was to eliminate many existing Apple products to free up resources for new products. Jobs actively participates in new product development and is the face of the firm when it comes time to announce new innovations. He embraces the new products and services from inception to product launch.

Relentless innovators demonstrate an open channel of communication between executives and middle management, and those executives understand the importance of an operating model balanced between efficiency and innovation. Executives in such firms aren't antagonists for innovation, they are protagonists, unified with middle management, actively working to encourage more innovation and build the structures necessary to sustain efforts in the long run.

Executives and their attitudes and behaviors matter to the success or failure of innovation. Executives cannot simply decree that they want more innovation. They must become far more specific about the new products and services they expect, must communicate their vision, help establish and build innovation skills and competencies, and participate in innovation activities. Their active involvement and commitment will demonstrate that the focus will be sustained, and that the business-as-usual model must be adjusted to accommodate innovation. Otherwise, the model will resist short-term innovation attempts, which will inevitably result in a crisis.

Chapter 6

The Usual Suspects

In a corporate setting, a crisis unfolds when an executive demands new products and services from unprepared, untrained, overworked middle managers. When executives demand radical or disruptive innovation from such individuals, managers are confronted with a stark reality: they must define and enact an activity that leads to innovative outcomes. Desperate times often call for desperate measures, and this confrontation is no exception. Managers who don't possess the skills, knowledge, bandwidth, or scope to innovate effectively must do so very quickly, without disrupting business as usual. Most organizations don't have the infrastructure or capability to innovate on demand; there are few people with innovation experience, little successful history to use as a guide, and no established methods or frameworks.

Since the organization isn't prepared and fully capable for innovation, middle managers must find a way to generate and manage innovative ideas and transition those ideas to new products or services while continuing to achieve consistent quarterly results. To accomplish these goals, middle managers often turn to methods, tools, or techniques that, in themselves, aren't unreasonable, but often aren't appropriate for rapid, disruptive innovation or to develop a long-term culture of innovation.

Middle managers generally seek to

- Minimize the impact of the request for innovation, by isolating the innovative efforts from business as usual
- Distribute the innovation work broadly (using idea management software), or
- Outsource the work entirely

Each of these reactions is completely understandable, because they represent potentially viable alternatives to developing an innovation discipline internally. They don't, however, change the attitude of the BAU toward innovation. These techniques seek to work around or on top of the existing BAU process. While these alternatives are initially attractive, they often become innovation dead ends unless the operating model is changed as well.

Isolation or Skunkworks

In a firm that doesn't have an innovation business discipline or an ongoing set of innovation processes and capabilities, a request for innovation will be interpreted as a one-time innovation project. If the middle managers accept that the request is important and relevant, their first inclination will be to isolate the innovation effort from the business-as-usual operating model, ensuring the continuity of the existing processes and improving the chances of achieving quarterly results. In effect, they protect the BAU operating model by isolating the innovation work.

To isolate the innovation effort, the middle managers will define a small team to work on the innovation effort in addition to their regular jobs. Typically, this team will work in some secrecy, physically or geographically isolated from the rest of the business, with as little public communication about their efforts and goals as possible. As mentioned, this approach to innovation is often called a "skunkworks," indicating a small team set aside to work on disruptive ideas and innovations. The term "skunkworks" originated at Lockheed Martin and it has been used to identify secretive, isolated project teams working on important projects. The first "skunkworks" was developed to design and build a new jet fighter during World War II,[1] and the concept has been used repeatedly by firms to develop radical new products or services.

Skunkworks traditionally have two purposes: one, to provide secrecy about the new product or service from the competition and two, to isolate the development of new ideas from the constraints and bureaucracy of the existing products and services. When used as intended, a skunkworks can be a powerful innovation platform. However, in many cases, middle managers use a skunkworks approach not to protect the new development from the existing BAU, but to protect the BAU from radical new concepts and ideas. Instead of accelerating new innovations to market, many skunkworks are often "make work" projects.

If the skunkworks team is working on innovation as a project in addition to their regular assignments, they meet infrequently and lack necessary skills. They work in isolation and they are encouraged to avoid any disruption to existing business processes, so they refrain from contacting people who have more insight or knowledge about specific needs or business processes. Their work remains cloaked in secrecy and eventually starts to spark questions about purpose and intent from the rest of the business. The limitations of this approach quickly become apparent.

> While skunkworks have been used to great success, their implementation is often intended to isolate disruptive ideas from the important work of the existing operating model.

Isolated from others, working on a part-time basis with little support and with the admonition not to impact any existing products or services, the team generates fanciful ideas that can't be delivered and incremental ideas that don't seem all that valuable. The result is unfortunate, because a well-managed skunkworks can generate truly market-disrupting ideas—"game changers"—but only if the teams are well-trained, well-led, well-supplied, and well-informed. In many cases, however, the skunkworks is used not to accelerate disruptive ideas but to eliminate any disruption to regular business by innovative ideas.

Skunkworks can be an effective tool within a larger innovation framework, especially when the innovation goal is radical, but they should be just one tool or technique within a range of innovation options, not the only approach to innovation. Further, a viable skunkworks operation requires good people who receive adequate training and leadership, and who aren't pressured to produce immediate results. Too often skunkworks are used as a stop-gap approach for "quick and dirty" innovation needs, but this is both a misuse of the concept and a poor substitute for the investment necessary for successful innovation.

Distributing the Work Through Idea Management Software

Another fairly typical response to an executive's demand for innovation is to acquire or develop an *idea management software solution*. Idea management solutions are simply software applications that assist with the generation, capture, and management of innovative ideas. Software typically becomes valuable in an innovation setting when

- There are a lot of ideas
- There are a lot of people participating in the activity, or
- The people involved in the innovation work are distributed geographically

Idea management software is appealing to many executives who are trying to introduce innovation, for several reasons:

- Software is a familiar solution. Software has been introduced to solve other problems in the business, and other firms have had some success with idea management software.
- The belief that innovation is simply a collection and management issue and all that is needed is a "place" for ideas to be captured.
- Since everyone is busy, distributing the innovation work means the firm can tap into more people to generate and manage ideas.

Executives are comfortable with the idea that enterprise software can reduce costs and improve efficiency. Idea management software appeals to this efficiency and cost-cutting bias, and software can provide excellent benefits when combined with a well-defined innovation process and trained personnel. However, idea management software by itself won't accelerate innovation. While the concept of idea management software seems promising, several challenges exist to this approach.

First, the organization must determine whether to "build or buy" the software. Given the exceptionally limited information technology resources in most firms, it is often much faster to license software from a third party rather than build it in house. However, in many organizations it takes months to evaluate and acquire idea management software from a third-party vendor, delaying an innovation initiative. Careful evaluation of idea management software is important to managers who will use the application as well as to internal IT teams. Managers must evaluate the software to ensure it meets the needs of the organization and is easy to learn and to use.

Internal information technology professionals must understand how the software works, how it is developed and supported, and they must give the software its blessing, even if it is licensed from a third-party vendor. Functional managers will also rely on the judgment of IT personnel to assess how secure an idea management application is, since the ideas can be considered intellectual property and should remain protected from other organizations. Further, negotiating a license and setting up a new system always take more time than anticipated. Even after the software is acquired, firms struggle to frame the most pertinent issues for the staff and managers, so the software can become a glorified suggestion box, filled with many ideas that have little relevance to the business.

Even when the software is implemented correctly and the innovation team defines *valuable* idea campaigns, many firms find that the software does little to speed the evaluation and conversion of ideas to new products and services. While ideas do benefit from broad interaction and participation by a large number of people, many can only be evaluated by a select group of people with deep subject matter expertise in specific fields, such as legal, regulatory, engineering, or sales. Idea evaluation, prioritization, and selection must be trusted to a smaller team with deeper skills and insights. Also, while a vast majority of the firm may believe that an idea is interesting and valid, at least one executive must agree to sponsor the idea and fund its conversion into a new product or service. Software can accelerate and distribute the work of generating ideas, and to some extent the ranking or prioritization of them, but many tasks, especially those related to review and evaluation of ideas and the selection of ideas for development, depend far more on well-trained staff working in a defined workflow.

Finally, idea management software doesn't resolve the issues confronting any innovation effort in regard to the BAU: too little time, focus, and strategy, accompanied by too much risk.

> Idea management software often highlights gaps in the innovation process or the lack of defined roles and responsibilities in the innovation process.

The acquisition and use of idea management software does not hinder the development and management of ideas, but often it simply highlights weaknesses or gaps farther downstream in the innovation process. Regardless of the number of ideas a team generates or how many people participate, ideas must be evaluated, selected, tested, and prototyped for their impact on the market. They also must be funded and commercialized to become new products and services. While you can support a good innovation process with idea management software, your team can't simply "automate" the innovation process and eliminate the need for well-trained people and defined processes.

Outsourcing Innovation

Another approach that many firms pursue in regards to innovation is to turn over idea generation, management, evaluation, and development to third-party consultants. This approach reduces the need to develop internal innovation methods, skills, and processes and reduces distraction to the existing "operating model." This option also benefits from new perspectives from individuals who are not bound by corporate expectations and culture. Third-party consultants can also offer more speed when compared to an internal team that isn't experienced, so executives often consider this option. There are, however, several significant challenges to using third-party idea management consultants.

First, consulting is expensive, and while innovation is rarely budgeted internally, internal innovation development is far *less* expensive than working with third parties. Innovation consulting can be even more expensive, for several reasons. There are few firms that span the innovation process—most "innovation" consultants focus on only one portion. For example, some focus on corporate strategy and the important "white space" areas for innovation; others are involved in trend spotting, while still others focus on gathering customer needs and insights. All of these skills are valuable, but few firms span all of these skills effectively. That means it's not unusual to work with several firms over the course of an innovation effort. Also, many of the firms that offer these services work in a traditional management consulting model, applying large and expensive teams to each challenge. These teams are necessary because executives don't appreciate the value of training their internal staff and developing their innovation skills, so few resources are available to work with the consulting teams.

Second, unless executives offer clear strategy and scope, it becomes easy for the consultants to deliver interesting ideas that aren't valuable or relevant for the business, or ideas that mimic the products and services of industry leaders. As we've discussed previously, innovation works most effectively when executives provide clear strategy, goals, and scope. Whether a team is composed of internal resources, a mixture of internal resources and consultants, or purely external consultants, the lack of strategy will stymie any team. Consultants feel more pressure to produce results than internal staff, so they will produce a report, a strategy, or a set of ideas. They are, however, just as limited by the absence of strategy as the internal staff. Additionally, since consultants work with a wide array of customers, it is possible that ideas that are similar to products and services already in the market may be recommended to executives.

Even if the consultants are affordable and deliver excellent ideas, those ideas still must be converted into new products or services in the development pipeline—the stage in which many good ideas get lost in transition. The idea must be developed by the product or service development team within an organization, then commercialized and launched effectively. Unless the consulting organization is tightly integrated with the product development and launch team and understands the capabilities and priorities of the development team, great ideas may never reach the market. This issue, again, is one that internal innovation teams face as well, but in most instances development teams have some awareness of the projects internal teams are developing. An outsourced consulting project, however, may take the development team by surprise. In this case, even if the idea is exceptionally valuable, there may be no way to quickly develop and commercialize it.

> The most significant hurdle ideas face in the route to becoming products is the transition from idea to funded product or service development. Only careful transition planning and active sponsorship of the ideas will carry them across this chasm.

This gap in the innovation process, between the selection of good ideas and the development of a new product or service, is perhaps the most significant hurdle faced in transforming ideas into products. No software or third-party consultant can solve this issue—what is needed is a well-defined transition plan that helps prioritize and allocate the resources and funds involved.

Finally, when working with third-party consultants, the firm rarely gains any knowledge or insight about innovation efforts. Instead, staff and managers push all of the innovation duties off onto the third party, never involving themselves in the process. If the internal innovators aren't able to commit time to the innovation effort, they typically don't receive any training and they aren't present in the development of ideas. Rather than build up its capabilities and knowledge, the firm becomes reliant on the consultants, turning to them again and again.

Internal resources are tasked with maintaining the BAU and possibly developing small, incremental innovation while third-party consultants develop more radical innovations. This reality explains why Six Sigma and Lean have been adopted as "innovation" tools by internal staff. They are simply deploying tools they understand to attempt to create incremental innovation, while leaving all new, radical, and disruptive innovation to third parties.

I've listed a number of concerns associated with innovation consultants, but I don't want to leave the wrong impression, since OVO offers innovation consulting services, and I make my living doing this work. Innovation consultants can create tremendous value, but many firms fail to tap the incredible resources within their own organization, generating anger and frustration on the part of the people who have good ideas and want to innovate. Rather than outsourcing innovation, firms should partner with innovation consultants who can offer to transfer insights and knowledge to internal teams, or who offer specialized skills. Increasingly, innovation needs to become a corporate business discipline, which means many of these skills must reside inside a firm. Corporations should work with innovation consultants to gain the benefits of a true "consultant" experience—project management, deep skills, and knowledge transfer—rather than simply outsourcing innovation.

Acknowledging the Short Cuts and Accelerating the Adoption

When pursuing these alternatives to direct, internal innovation, executives who want more innovation may determine that it isn't possible or the organization isn't responsive enough to their needs. In that case the executive will shift focus to greater internal productivity or seek to acquire a solution externally.

Another possible outcome is that the executive team confronts the "elephant in the room" and develops innovation as a business discipline.

Attempting to bypass or shortcut business as usual through skunkworks and software or relying on outsourced idea consultants may speed part of the innovation process, but these actions don't account for all the activity and commitment necessary to convert an idea into a new product or service. Eventually, when some or all these efforts have been attempted, the "crisis" will have to be confronted head on—the BAU process and the barriers it creates to consistent innovation must be re-evaluated.

It's difficult for senior executives to acknowledge the truth that their firm simply isn't organized or constructed to innovate consistently. It's with this recognition, however, that important changes within the firm can begin. Such a transformation can only occur if there is acceptance at the senior executive levels of innovation as an important competitive advantage that firms must exercise consistently over time, with one common approach and method. Just as corporate culture and common business processes are developed and improved, an innovation capability will take time and discipline to create. Innovation will require a consistent "workflow," since people work most effectively when they understand the work and their place or role in the work.

> Only when executives recognize that innovation is a competitive advantage and the firm must innovate consistently over time with a defined approach will the operating model change.

Additionally, roles and responsibilities must be defined for the innovation process so that in every step of the innovation activity people understand their roles and add value to the process. Roles and responsibilities must be defined and the people who fill those roles—on a full-time basis or part-time basis—must be trained. Further, their compensation and evaluations must change to reflect the importance of the innovation work they are called on to do. When executives realize the need for consistent innovation, and understand the investment necessary to achieve that innovation, then the BAU operating model will change.

A Core Innovation Team

There are several actions that can accelerate the transition from an operating model focused exclusively on efficiency to one balanced between efficiency and innovation. Probably the most important action is to create a *core innovation team.*

A core innovation team is responsible for defining the innovation processes, methods, tools, language, and culture, and it is vital to a firm's success. The team collects and shares the best processes and works with any product team, geographic team, or line of business to assist in an innovation effort. Just as everyone in the organization follows a consistent purchasing process, the core innovation team provides a consistent model and method for innovation, offering tools, techniques, and support to assist any group.

Innovation can occur anywhere in an organization—in a product line, in a business unit, or in specific geographies. Regardless of "where" innovation happens, it should be conducted with as consistent a model as possible to ensure that ideas can be compared across business units or product groups when requests for funding or resources are made. If every group within a firm follows a consistent innovation process and method, then it will be much easier for executive management to compare ideas, understand risks, and make funding decisions. As groups exercise the same models and adapt them for their uses, they improve their innovation skills over time.

All the following components that are necessary for innovation take time, resources and, most important, intent:

- Aligning compensation and culture to innovation goals
- Improving the innovation skills and methods of your employees
- Developing a central innovation method or process
- Implementing idea management software
- Developing a central innovation team

If your organization has reached the "crisis"—it has tried everything to innovate but it hasn't been ultimately successful, one or more of these items is missing, and innovation as a consistent capability is only possible when all of these factors work together effectively. Let's look at what it takes to create an innovation BAU culture.

Making the Transition

The difference between relentless innovators and your firm is that these innovators have faced the crisis head-on. They have transitioned to an operating model balanced between efficiency and innovation, not one that relies solely on "efficiency and effectiveness." After confronting the crisis they decided to create an innovation BAU operating model and corporate culture that sustains innovation *and* efficiency. Such relentless innovators now are able to consistently generate new ideas and commercialize products and services with great success.

What you'll also notice about these relentless innovators is that they use "skunkworks," idea management software, and third-party consultants effectively as part of their overall innovation strategy. These powerful, valuable tools and assets are just part of the innovation business-as-usual framework, rather than an attempt to layer innovation on a resistant business-as-usual operating model.

Let's look at how to confront the innovation crisis and build an innovation business-as-usual model.

Chapter 7

Creating an Innovation Business-as-Usual Approach

I've spent the last few chapters investigating why many firms can't innovate successfully, and I have identified two culprits that keep most organizations from doing so: middle management and the business-as-usual operating model. In the last chapter I looked at the typical but often inadequate steps that many firms attempt before committing to innovation as a discipline. Often it is only when these measures aren't fruitful that executives reassess their approach and commit their organizations to sustained innovation.

In this chapter I'll look at several firms that have sustained innovation success over many years, and consider how their BAU operating models allow them to embrace, rather than reject, innovation as part of their regular efforts. From this investigation I'll develop a framework that illuminates the attributes a business must adopt to shift the perspectives of middle managers and change the BAU process to an *innovation* BAU process.

The Big Question

The big question that has confronted us throughout this book is: how do successful innovators sustain innovation over a long period of time? Even though firms like 3M, Procter & Gamble, or Apple produce different products and exist in different markets, something about their cultures or models enables them to consistently generate new products and services, often over decades. What is it about their perspectives, cultures, and processes that allow these firms to sustain innovation rather than considering it a threat to business-as-usual processes? What can a business that wants to sustain innovation and create an internal innovation capability learn from these firms? Can *any* firm modify existing BAU processes to embrace innovation?

Many factors can *contribute* to a firm's innovation capability, including

- Charismatic leadership
- Dynamic, engaging cultures
- Deep research and development skills
- Close relationships with customers and partners

However, none of these factors sustain innovation over time, or are even required for innovation to succeed. Consider W. L. Gore.

Gore is an interesting corporate environment, a case study for organizational structure and behavior. As mentioned, it is the embodiment of a fully actualized employee organization, where employees elect their leaders and have great autonomy to pursue their ideas. Jim Collins investigated Gore's corporate culture in *Built to Last*, and Gary Hamel, another management strategist who has written about culture, leadership, and innovation, has explored Gore's organizational structures in *The Future of Management*. Gore demonstrates that in an organization with little hierarchy and little "top down" charismatic leadership, innovation can thrive. Contrast that philosophy with Apple, where most of the new ideas come from a small team of charismatic senior executives.

These attributes listed can, in some instances, accelerate or slow innovation efforts, but, as you should know by now, the primary driver of innovation success is a combination of culture, attitudes, frameworks, and processes that form the accepted "operating model" for the business.

What Factors Create an "Innovation Business-as-Usual" Framework?

Before we examine what factors sustain and enable consistent innovation, let's define the goal for this new innovation BAU organization. In a perfect world, an organization effectively balancing efficiency and innovation capabilities will continue to deliver quarterly results on a consistent basis. However, the perspectives on innovation will change dramatically. The goal for the operating model is to look something like this:

- The operating model will consider innovation a persistent capability and a business discipline, rather than as an occasional disruptive initiative or a discrete activity reacting to market conditions.
- Innovation is broadly defined, meant to seek out new ideas that can add value by cutting costs, improving efficiency internally, and creating interesting new products, services, and business models.
- Everyone—executives, staff, and especially middle managers—expect to innovate and accept innovation as a common business practice.
- Innovation will be tightly linked to strategy and the goals will be clearly communicated.
- The "operating model" for the business will reinforce both efficiency and innovation.
- New skills, techniques, and methods will also be introduced to speed effectiveness.
- The executive team and middle managers will constantly seek out new ideas.
- Well-defined innovation processes will move ideas to evaluation and selection.
- Executives will ensure a smooth transition from nascent idea to viable product by building bridges between idea management and product development processes.
- Commercialization and launch programs will support new ideas as they become products and services.

- Innovation is understood as a consistent business discipline reinforced by a well-defined innovation process from idea generation to market launch.
- Innovation becomes a "way of life" rather than an occasional disrupter to core processes.

Everyone—executives, staff, and middle managers—will arrive at work expecting to innovate and they will be well versed in the tools, processes, and methods of innovation.

What factors need to be changed or introduced to achieve this goal? Considering some of the recognized innovation leaders, it's easy to identify eight factors that create an innovation BAU framework in their organizations: (1) innovation metrics tied to specific strategic goals; (2) compensation; (3) enabling functions; (4) who we manage versus what we manage; (5) communication; (6) defined processes; (7) reactive versus proactive philosophy; and (8) human resources and talent management. Let's examine these eight factors and consider how to implement them to craft an innovation BAU operating model in your business.

Innovation Metrics Tied to Specific Strategic Goals

Executives often fail to link requests for innovation to specific supporting strategies or goals. Contrast this lack of clarity with probably the most ambitious innovation goal in the market in the first decade of the twenty-first century: Arthur Lafley's goal that 50 percent of Procter & Gamble's ideas would originate from outside the company.[1] For a company with a strong corporate culture and a large R&D staff that have been doing a reasonable job at creating new products, this goal must have come as quite a shock (it certainly did to the rest of the consumer packaged goods market). If Procter & Gamble, the leader in many consumer packaged goods segments with one of the largest R&D teams in the world, couldn't create enough new products and ideas internally to keep an innovation funnel filled, what firm could?

Lafley recognized that there were more needs and ideas in the world than his R&D teams could identify or pursue by themselves. Just establishing this goal forced the entire organization to consider how innovation would get done, since the goals were specific and measurable. This goal caused a significant shift toward "open innovation" for P&G that has paid significant benefits. In 2009, Procter & Gamble produced five of the top ten new product launches in the United States, and more than 60 percent of Procter & Gamble's new products were generated with at least one external partner. Products generated from Procter & Gamble's Connect + Develop program accounted for $1 billion in sales in 2009, just eight years after the program was launched.[2]

It is important to note that Lafley didn't merely set an internal goal, he established the goal *publicly*, where Procter & Gamble's partners, customers, and shareholders could hear it and understand it. By communicating it in such a way, he committed himself and his management team to specific measurable goals, which solidified the strategy for the rest of the organization.

Or consider 3M's goal that 30 percent of the revenue generated in any year should come from products that are less than four years old. 3M, therefore, can't rely on its base of excellent products, which drive substantial revenue, but must constantly create new products and services to fulfill that mission. This strategic goal with a well-defined measure focuses the entire organization on the task at hand.

A clear innovation goal, supported by a *quantifiable metric*, announced and sponsored by a senior executive, demonstrates to the organization the intention to develop and sustain innovation capabilities, and forces the BAU structures to shift. At this level, and with this focus, innovation clearly isn't a "one-time" event and it can't be walled-off from the rest of the organization. Existing processes and frameworks have to adjust to achieve goals, rather than altering ideas to fit existing processes and expectations.

Setting Fences. Clearly established innovation goals also define scope, resources, and a timeline. The *scope* establishes expectations about investments and timeframes. Denny Potter, the vice president of innovation at RJ Reynolds Tobacco Company, calls this process "setting fences." These "fences" can be thought of as innovation horizons, defining incremental change as "Horizon 1," breakthrough innovation as "Horizon 2," and radical, disruptive innovation as "Horizon 3." Executives must clearly delineate the innovation goals and investment in each of the three horizons or "fences." Without that delineation, innovators are likely to either constrain themselves completely within Horizon 1, neglecting longer term, disruptive innovation opportunities, or spend too much time in Horizon 3 at the expense of continuity, ignoring Horizon 2. Potter identified this problem and the need to establish innovation "fences": "...hindsight also shows us that we succeeded in a high rate of unconnected, short-term focused, incremental innovations with a few evolutionary and a very few break-through product offerings."[3]

A reasonable innovation effort should countenance innovation efforts in all the horizons, with strategic goals dictating the distribution of resources and time in the three horizons.

Cascading Expectations. The establishment of corporate goals has the additional benefit of cascading into product teams and lines of business. As an example from OVO's clientele, an executive vice president in charge of a major U.S. financial institution is requiring an annual plan from her direct reports defining the number of innovation activities their teams will undertake and the expected revenue impact of those activities for the year. By incorporating innovation as a defined task in an annual planning cycle, she has set the expectation that innovation is a sustainable capability that she expects each line of business to deploy several times in a fiscal cycle.

This expectation creates a cascading effect, requiring product teams and business line teams to define their innovation actions for the year, reinforcing the demand and importance of innovation. Further, it establishes the expectation that innovation activities will be conducted consistently and measured regularly. Since the innovation activities are part of an annual plan, innovation is now also part of her direct reports' compensation plan. Innovation activities will receive far greater attention when they are incorporated in the executive's plan to influence compensation.

> Executives must be specific about their innovation goals, they must link innovation to key strategies, and they must develop measures and metrics to hold innovators—and themselves—accountable.

Goals Must Be Tied to Profitability. Note what is important and consistent in all of these examples: a senior executive creates a specific innovation target for the business in terms of its contribution to growth, revenue, or profits and communicates that goal to everyone in the business. Ideally, those corporate goals are integrated into the business plans that the direct reports create, and they become part of the compensation and evaluation plans of senior executives. These actions create a clear, unmistakable message that innovation is important, with specific purposes and goals imperative to the business. Innovation will be effectively measured and rewarded.

In your organization you may not be able to establish an innovation directive from the CEO. Innovation may start as a goal within one line of business, one product group, or one geography. While "top down" innovation is optimal, root innovation deeply where you can achieve it by establishing clear goals and measurements at the highest level of the organization you can successfully influence. If your innovation program is isolated in one product group or geography, communicate the innovation goal within that team, and follow up with regular measurements and reports.

Choose goals and metrics that impact the revenue and profitability of your business, and preferably choose "stretch" goals that force your team to use all facets of innovation, rather than safe goals they can achieve through incremental change. Establish clear objectives and constantly reinforce those objectives to demonstrate that they are strategic, important to senior leadership, and not subject to the ups and downs of the economic cycle. Both 3M's and P&G's innovation goals are broadly recognized, they have been in place for several years, and they are measurable. Neither of their goals has altered in market declines.

Strawman Strategy. In the absence of clear strategy, executives and middle managers will place a disproportionate amount of emphasis on maintaining consistency and continuity rather than introducing innovation. When a clear strategic goal for innovation doesn't exist or isn't clearly communicated, the best course of action is to create a "strawman" strategy that defines innovation goals. This "strawman" is an attempt to define the goals of innovation and gain executive approval for an innovation initiative in the absence of clear strategy. The strawman allows innovators to define what they believe the strategy should be and circulate that strawman to executives. While this slows the innovation effort, it provides a much greater chance of success than merely plunging in with no clear targets or goals. In my experience, most innovation teams don't take this step and they are quickly left floundering because the scope of the effort and their targets or goals are so poorly defined.

Compensation

Compensation in many organizations is driven by three factors: the salary bands or grades that an employee is assigned to, the outcomes of the business as a whole, and the evaluation of the employee's work in the previous year. People want to do work that engages their interests and passions, but they must work to optimize their compensation. While their hearts may be aligned to innovation activities, their minds and wallets are focused on how they'll eventually be evaluated and compensated. This means that the operating model must introduce new evaluation metrics, incorporating more weight on innovation activities and leading to a balanced outcome for compensation, advancement, and promotion.

Humans are rational actors who seek to undertake work resulting in rewards and avoiding actions that lead to reprimands. In large firms, human resource teams spend countless hours developing evaluation and compensation schemes meant to ensure that employees are adequately and fairly compensated for their work, while being encouraged and rewarded for activities aligning to the success of the business. Due to the strong evaluation and compensation programs in place in many businesses, and the overriding focus on efficiency and effectiveness, innovation is difficult to sustain—innovators are often assigned to an innovation role on a part-time basis but their compensation and advancement remains tied solely to the evaluations of the work they accomplish on their "day jobs." When push comes to shove, employees' focus, attention, and effort will revert to their regular duties.

Therefore, to sustain an innovation focus in a business, organizations must change their evaluation and compensation structures. Few, if any, organizations have well-defined criteria to evaluate work associated with innovation. Since compensation is highly correlated to evaluation results, if the evaluation program doesn't measure or recognize innovation efforts, there will be little additional compensation for innovation activities, encouraging employees to focus on business as usual over innovation. If the compensation plan rewards business as usual and snubs or omits rewards or compensation for innovation, why should any manager focus on innovation over business as usual?

Linking Innovation Projects with Evaluation. OVO's banking client went further than just requiring innovation projects in the annual plan—they now link innovation projects to the evaluation and compensation for senior executives in their business lines. Executives have an added incentive to be innovative—their evaluations, promotions, and compensation are directly impacted by their innovation efforts. There's an obvious importance in innovation for these executives, when clarity about compensation is often lacking for many potential innovators. Google, as another example, offers innovators a stake in the rewards of the ideas that are converted into products or services, so the individuals see a direct result of their efforts in their paychecks or stock awards.

> Lower the barriers that keep people from getting involved. Develop evaluation plans and compensation models to incentivize your best people to become innovation leaders and sponsors.

Encourage and Reward. Rather than allow compensation to dissuade strong individuals from innovation work, the new innovation BAU operating model demands compensation schemes that encourage and reward people who innovate. Both the compensation and evaluation schemes must demonstrate that a focus on innovation will be rewarded.

Compensation programs also must account for the fact that there are different roles in an innovation effort. A few new full-time roles may be created to support innovation, while many individuals in the organization may be called on to participate on a part-time basis as idea generators, evaluators, or subject matter experts. These innovation activities need to receive as much "weight" in the evaluation and compensation program as other tasks that sustain business as usual. Further, gaining new experience that supports innovation, whether through classroom training or actively participating in an innovation effort, must be recognized and rewarded as well.

Enabling Functions

When thinking about innovation, it's easy to assume that idea generation and product development are the only important activities or functions. However, in any large enterprise there are important teams that are required to evaluate, modify, and approve a product or service that are often overlooked or ignored.

Business functions such as legal, regulatory, information technology, and compliance can delay or block the development of a new product or service, hamper the launch of a new product or service, or fail to support the product or service in the marketplace. I've labeled these functions and teams as "enabling" functions. They may or may not play an active role in the development of an idea, but they take on an important role as an idea matures into a new product or service. They can also have powerful *adverse* effects on an idea late in the development cycle. In fact, these teams are often the scapegoats for innovation failure, for at least three reasons.

First, innovation teams don't always alert these enabling functions to new ideas early enough in the process. The enabling teams need to provide feedback that will ensure the ideas align to legal or regulatory constraints or that can be supported in the market in advance. If enabling functions are not introduced to the idea early on, innovations are often blocked late in the development process due to problems that could have potentially been avoided.

Second, employees in these enabling functions often have exceptionally limited resources and they react negatively to any unanticipated change or request for new resources. For example, the vast majority of the information technology team's budget is dedicated to maintenance of existing hardware and software. If your innovation requires new software or even modifications to existing software applications for success, you may find the IT team unable to provide the services and support you need.

Third, many of these enabling services provide "protective" functions, working to protect the firm from producing dangerous products that may cause harm to consumers in the market. It is incumbent on these protective functions to subject the new product or service to close scrutiny. Legal, regulatory, and compliance teams may need a significant amount of time to review a new product or service, and they may demand significant changes to ideas late in the development process.

Enabling Functions in the Innovation Framework. All of these "enabling functions" have valid reasons to reject innovative ideas, so any shift in the operating model must also incorporate changes to the way enabling functions are involved. To bring these teams into the new innovation BAU framework, your organization must address these factors:

- **The expectation and metrics of the enabling functions.** Currently these functions are constrained by resources and budgets (information technology) or by the perspective their responsibilities place on them (protective functions like compliance and regulatory). In both instances, these functions are gatekeepers and potential barriers for new products and services. In order for your new innovation BAU process to be as effective as possible, your firm must reset the perspectives and revisit the roles of these enabling and protective functions. These functions must adopt a proactive, supportive position on new products and services, seeking to advance as many as possible, as quickly as possible, rather than a reactive, defensive posture from which it is easy to reject new ideas.
- **Compensation and rewards.** Many enabling functions exist to protect the firm and to help it operate efficiently, and compensation is based on these goals. Yet these functions play a significant role in supporting and launching new ideas as products and services. Just

as the operating model must become more balanced for innovation success, the motivations and compensation for these enabling teams must be balanced between their protective functions and their ability to support and enable new products in the market.

- **Early introduction of the enabling functions to innovative ideas.** Given the enabling functions' role, and the expectations of business and compensation programs, employees performing these functions are far more likely to reject new ideas than to change existing operating models. The potential for rejection or rework can be dramatically reduced by incorporating these enabling function team members on innovation efforts as early as possible, and keeping the functions well informed of valuable new ideas. In this way the enabling functions can help shape the idea to achieve approval with fewer changes and rejections.

You'll note that many of the recommendations made here are similar to the recommendations made more broadly about the changes necessary for innovation as a whole. I've broken out these suggestions about the enabling functions—IT, legal, regulatory, compliance, and so forth—because they are rarely considered as part of the innovation ecosystem until it is too late to modify the ideas, products, and services. While they are only tangentially involved in innovation, the small role they play is vitally important. Shifting the operating model to an innovation BAU model won't work unless these enabling functions are incorporated into the change, as much a part of it as the core innovation teams and processes.

Who We Manage versus What We Manage

As noted previously, many innovation projects have difficulty attracting top talent for several reasons:

- The best people in an organization are already in high demand, working on mission-critical tasks.
- Most people already have more than a full workload, so it is difficult to find good executives or managers with spare "bandwidth" for an innovation activity.
- It's hard to assign valuable executives to what is often considered a risky innovation effort—their skills are vital to achieving important short-term objectives.
- Given the choice, many experienced managers would prefer to forego the risks associated with innovation and stick with tried and true initiatives.
- It's hard to find good leadership of an innovation initiative because the likelihood of failure is high, and talented managers and executives don't want a taint of failure on their performance record.

All of these reasons, however, pale in light of this: the size and scope of the project is relatively small in an era when managing large teams or budgets demonstrates seniority.

Seniority and Status. In many organizations, how many people you manage is a sign of your status and importance to the organization: the larger the pyramid, the more important the role. With a larger pyramid comes greater responsibility and compensation. Few executives are willing to move from a position that manages a large number of people to one that manages a small group, but that's exactly what needs to happen in an innovation project.

The best people in your organization are attracted to the leadership opportunities that provide the best option for growth and advancement. The best, most experienced people manage the "core" products and functions and other junior or less experienced managers oversee innovation. In an innovation BAU firm, these allotments are more balanced. In fact *who you manage*, or how *large your pyramid* is, needs to be balanced against *the value of the ideas that you manage.*

Wouldn't it be interesting if the most compelling management role in an organization was to lead an innovation project? After all, if executives believe innovation is so important, shouldn't they place their top people on innovation efforts?

> Wouldn't it be interesting if the most compelling and in-demand role in an organization was leadership of an innovation project?

IBM faced this issue and decided to utilize its best people in emerging growth opportunities. Here's the story, with direct reference to a Stanford Business school paper covering the topic.[4]

IBM Example. In the late 1990s, IBM CEO Lou Gerstner learned that financial pressures had forced an IBM business unit to discontinue funding of a new initiative. He demanded to know why the company failed to identify and fund emerging opportunities. There were several key responses that will seem familiar:

- Short-term execution focus
- Management rewards
- A focus on exploiting "known" markets
- Inadequate insight to embryonic markets
- No established processes for experimenting and growing new businesses

In response to the failure to identify and utilize new markets, IBM set up its emerging business organization (EBO) to exploit new opportunities. One of the most important changes they made with the EBO was to decide how these emerging opportunities should be led. Here I quote directly from the paper:

> Historically, when IBM chose leaders for new growth initiatives, the tendency was to select younger, less experienced people to manage the projects. The logic was that younger leaders would be less imbued with the "IBM way" and more likely to try new approaches. These leaders often failed. What the company learned was younger managers often lacked the networks needed to nurture an embryonic business within the larger company. . . .[5]
>
> Rod Adkins was a star within the company who was running the thriving UNIX business with 35,000 employees and $4 billion in sales. When he was chosen in 2000 to run the new pervasive computing EBO, a business with zero revenues, his first thought was that he had been fired. . . .[6]
>
> Over time, the success of the EBO effort has made running an EBO a desirable job, with people volunteering to run them.[7]

While the EBOs are not an exact corollary to a new innovation initiative, the same issues and challenges are presented in both instances. IBM decided to place its best people on emerging opportunities and the results have been exceptional, both for the firm and for the individuals who led the EBOs. Those leadership roles, once considered dangerous, now have more volunteers than management slots. Putting your best people on your most important growth opportunities is a risky decision, but one that can pay significant dividends in the end.

Risk is a vital barrier to attracting good managers and team members as well. Business cultures have defined "failure" as a career limiting outcome, regardless of the circumstances, so few employees want to be involved in a project that has a high likelihood of failure—or even one that may struggle to success. This situation makes creating a new innovation initiative difficult, since everyone understands the risks, uncertainties, and false starts that are likely in an innovation effort. To attract strong people to leadership positions and innovation teams, firms must reduce the fear and uncertainty that surrounds any potential "failure" of an innovation project. The best way to do so is by incorporating the learning from the "failure" into a new effort and demonstrating that fast prototyping, experimentation, and rapid learning is a valuable part of the innovation experience. As long as the potential for "failure" exists in an innovation setting, however, your best people will avoid innovation.

Project Leaders' Capabilities. Not every person is capable of leading an innovation project. Just as there are people who naturally gravitate to sales leadership or product leadership or geographic leadership, there may be people who are best deployed in an innovation leadership role. Perhaps your best outcome is to identify and use these individuals in successive innovation projects, sponsored by line executives or product group heads. In such a difficult but important role, developing people who have the aptitude and skills to lead high-risk, high-profile projects can be exceptionally valuable. Leveraging those skills repeatedly is far better than requiring individuals to run an innovation effort who don't have the necessary abilities or interest. In the example of the IBM EBO, experienced managers and executives weren't simply parachuted in to these new EBOs; they were "trained in the skills needed for the emerging opportunities. The challenge, unlike in mature businesses, is not to empire build and staff up quickly but to get strategic clarity."[8]

Using Your Best People. Corporations need to place their best managers and leaders as innovation leaders and their best employees as innovation team members. The necessary shift will be to place emphasis on the ideas an individual or team manages rather than on the size or shape of the pyramid they manage. W. L. Gore is a good example of this thinking.

Gore has a rule of thumb that suggests any product line or business unit larger than 150 people is difficult to manage, creating bureaucracy rather than adding value. Gore intentionally keeps its product teams small and encourages the development of new teams when new ideas or products are generated.

To attract the best people to innovation challenges, firms must reduce both the risk associated with failure and the status and power associated with managing a large pyramid. A firm will know innovation has become business as usual when the strongest, most talented people vie for innovation roles, rather than avoid them.

Google is also a good example of building teams with the right people and the right passion. At Google, anyone can create an idea, but the ideas that are valued are those that attract team members willing to commit their time and resources to further the innovation. In a setting like Google's, only the best ideas will attract others. Those who support the idea also demonstrate passion and engagement for the innovation.

These two words, *passion* and *engagement*, are what your firm should strive for when building an innovation team. Finding strong people who have passion for solving a problem or filling a customer need while being fully engaged means that the team is much more likely to overcome challenges and barriers.

To change the focus of your "operating model" you'll need strong, competent executives and managers to lead innovation efforts. It will be difficult to attract the best leaders unless they understand how critically important innovation success is to the business. Further, once the importance of innovation success and leadership are recognized, many will demand more training, more preparation, and more support. Developing innovative leaders is no more difficult than developing efficient managers who deliver consistent results. Instead of spending an extraordinary amount of resources and time focused on developing skills and attributes to improve efficiency, many businesses will now need to spend time developing their workforce's innovation skills and capabilities.

Communication

Good communication is vitally important, and often exceptionally difficult, in a modern business. Difficulties arise from a number of issues, one being that many employees feel overwhelmed by the vast array of communication they receive. Due to the uncertainty and inherent risk of new innovations, clear communication of messages, intent, and the goal of innovation efforts is necessary throughout the organization and to external customers, partners, and markets. Innovators understand the role communication plays in the process and therefore focus on at least three important types.

The first type of communication is about the strategic nature of innovation. This communication flows throughout the organization from the top down and it is focused on purpose, goals, and intent. This information must trickle down through middle managers to all levels of the organization. The channels and methods you use are less important than the consistency of the messaging and the follow-up to demonstrate that the organization is measuring and reporting innovation goals and metrics. Explanation of these strategies reinforces the purpose and goals of innovation and demonstrates an ongoing management commitment to innovation. We've demonstrated two excellent examples of this kind of communication, Lafley's announcement of P&G's open innovation goal and 3M's stated intent that 30 percent of its profits must come from products that are four years old or newer. These communications from Lafley and from 3M are used extensively to demonstrate a public commitment to a quantifiable innovation metric because they are succinct and measurable, but also because they are so unusual. A scan of the business press reveals very few clear, succinct, and quantifiable communications about innovation intent. These communications align innovation to strategic goals and demonstrate the intent and commitment of senior executives to innovation outcomes.

Another type of communication relates to structure and processes. These internal communications are meant to initiate and sustain change within the organization and they are targeted at innovators and other individuals and teams who are tangentially involved in innovation (e.g., finance, legal, human resources). The goal of this communication is to reinforce the strategies and begin to implement the processes, methods, and techniques necessary for sustained innovation through the transition from a purely efficiency focused operating model to one balanced between efficiency and innovation.

The third type of communication necessary for sustained innovation is external, targeting existing customers, prospects, and the market. These communications alert your channels and customers about new methods, products, and intentions. Keeping these segments informed of your company's actions not only result in greater engagement, but can also lead to sales and increased profits down the line.

Once the communications that set the stage for innovation are underway, another type of communication is necessary—to the innovators and their internal communities. Having established the intent, goals, and commitment for innovation from the top down, executives now pass the communication baton to innovators, executives, and managers who begin to communicate not the "why" but the "how" of the effort. This tactical communication lays the groundwork for the changes that must take place in order for innovation to succeed. The impending changes, and the rationale for them, are described to those directly affected (such as the innovators, product development teams, and marketing) and to those indirectly affected (legal, human resources). These communications establish how work will get done in the new innovation BAU framework and link the innovation efforts back to the strategic goals established by executives.

Once the innovation capabilities and processes are in place and ready to generate and release ideas, another level of communication should occur. These communications target customers, prospects, and partners, focused on introducing new methods, such as open innovation, and new ideas as new products and services. Not only does the firm need new interaction models, but it needs to consider how it communicates its new products and services and how it *launches* those new products, which is also a function of communication.

Surprising the Market. Good innovators are recognized for their commitment to innovation. No one is surprised that they innovate. Consumers aren't shocked when Google brings out a new browser or operating system or surprised when P&G introduces new consumer goods. Given all the focus on innovation, your firm isn't going to surprise anyone by simply conducting an innovation initiative. In most cases, it will be evident to your customers and to your competitors that you are pursuing innovation goals. However, even though others may be aware that you are innovating, you may surprise or disrupt a market with the ideas you create—if your goal is radical innovation, that's a desired outcome.

> No one is surprised when innovation leaders create new products or services. We expect that. But customers are often pleasantly surprised by what the leaders produce.

Good innovators, however, often surprise the markets by *what* they produce. While their strategic focus on innovation is well communicated, internally and externally, they understand how to unveil their ideas—what to share and what to keep under wraps. Apple is a master at understanding how to communicate in this way. We expect Apple to innovate, and communication sets the stage for Apple, for its employees, customers, and competitors. Yet we are often pleasantly surprised by what Apple creates, because the company understands what to communicate, which channels to use, and how much information to make available to a waiting public.

Marketing and Positioning. Your communication programs also need to become more attuned to your marketing and positioning. Many good innovators understand how to use their innovative ideas, products, and solutions to gain more publicity. Of course, a lot of that publicity is based on the fact that the leading innovators discussed previously create interesting, valuable, and disruptive new products and services. If all of your ideas are incremental, it's hard to gain much publicity. Using your ideas as part of your marketing and communications campaign is valuable, especially if your organization is consistently innovating over time.

Lack of Communication. In many instances, an innovation initiative is considered an experiment, a small project that an executive feels inclined to test. In these situations, there may seem to be little advantage to communicating much about the effort to the rest of the organization—there is a high probability the effort won't be successful and if it is not well-known throughout the firm, once it fails it can be shut down and quietly swept under the rug. The lack of communication in itself, however, is a communication, saying to the rest of the organization that the initiative isn't strategic and it won't be sustained.

Contrast the quiet "sweep it under the rug" communication approach to what Lafley did at P&G, when he announced P&G's open innovation goals in the general business press. Once Lafley made that 50 percent commitment in such a public communication, he and his management team had no choice but to staff the required positions and fund the efforts. His public communication signaled his commitment to innovation to the organization, and his expectation of their efforts and commitment. Note as well the difference between Lafley's communication and what we've termed earlier a "flavor of the month." Serious strategy starts at the top, it aligns to corporate goals, and it has a clearly established objective that will be measured. Middle managers are always careful to distinguish between a carelessly tossed out demand and a specific request backed by resources that is carefully scoped.

Communication is a form of endorsement. Having an executive who talks to Wall Street or the financial community about innovation is commonplace. Read any annual report and you'll see the term "innovation" tossed around dozens of times by firms that, in reality, have little engagement with true innovation. Having a CEO or executive establish firm, public goals about innovation endorses the effort and reinforces the commitment. No one can be uncertain or unclear about Lafley's expectations, and no one can claim they didn't understand the urgency or importance of open innovation. However, when innovation projects are swept under the rug, not acknowledged, and poorly communicated, executive management makes a different endorsement, saying they don't support the innovation.

That lack of support in turn sends a clear message to the rest of the organization about their commitment levels to an innovation effort.

Defined Processes

Perhaps one of the biggest myths about innovation is the idea of the "lone" innovator, who works on ideas in the lab or office, without assistance or support. In this myth the innovator or inventor has a flash of insight, generates and manages ideas completely on his or her own, and fights the bureaucracy to overcome all odds to produce a commercially viable product. While these stories about individual innovators overcoming all odds are enjoyable, they are rarely true. In fact most, if not all, ideas that become new products or services require the involvement of a significant number of people from a wide array of business functions—sales, marketing, legal, manufacturing, and distribution, to name a few. The complexity inherent in developing, testing, and commercializing a new product demands a broad perspective and a diverse set of skills.

Likewise, innovators need strong, consistent processes and frameworks in order to manage, develop, and test ideas. Few firms succeed using ad hoc or "on the fly" innovation processes. A well-defined idea management and development process assists an innovator by reducing complexity, defining evaluation criteria, establishing "gates" and reviews for the ideas, and communicating workflow and tasks for the people who are involved in developing and managing ideas. A common, consistent process increases effectiveness, reduces bias in idea consideration, and encourages the development of institutional capabilities over time. When many teams or individuals attempt innovation using the same processes and methods, learning benefits become evident, reducing risks, costs, and timeframes associated with innovation and producing better results than in organizations that fail to define and sustain an innovation process.

A well-defined innovation process will encompass an entire "end to end" innovation capability, including these phases:

- Trend spotting and scenario planning
- Gathering customer needs and market insights
- Generating ideas using the scenarios and needs as guideposts
- Evaluating, prioritizing, and selecting ideas for further development
- Prototyping and piloting ideas
- Transitioning ideas into product or service development
- Launching new products and services

In each of these phases, there are a number of steps to complete the phase successfully. Further, each phase has a number of tools and techniques that must be mastered in order to produce effective results. To implement those tools and techniques and to complete this process, clearly defined workflow must exist, and the people who are expected to do this work must be trained. The innovation process is similar to other business processes within your firm. There must be a clear definition of the work, who does the work in each step, and a carefully defined workflow so that teams in each phase or activity understand the results of the work upstream and they can use that input to accomplish their tasks.

While innovation is consistently ranked as one of the most important capabilities, few firms have well-defined innovation processes or capabilities. Other important business processes, such as receiving customer orders or accepting payment for customer orders, are well-defined processes honed over years or decades. Yet innovation is still relegated in many firms to an ad hoc process developed by the innovator or innovation team, purpose-built for the task at hand, and rarely reused or repeated. No other important process is conducted in such an ad hoc manner. Innovation needs and deserves the same definition and process that other important functions benefit from.

> If innovation is so important in your business, why does your firm insist on an ad hoc innovation process? No other important function in your business is ad hoc.

When a Process Is Valuable. Defining and developing an innovation process, however, only makes sense if the process will be repeated. If an innovation initiative is a "once and done" event, developing a new innovation process specifically for a discrete, one-time initiative will not be worth the effort. Since few firms think of innovation as a business discipline that can be sustained over time, it doesn't seem useful to construct a consistent, repeatable innovation process, especially one that encompasses all of the tasks and phases identified above.

Further, defining a new process requires identifying roles and responsibilities to support and sustain the process, meaning that roles are created and education and training is required. Building an innovation process and staffing it effectively, though, isn't valuable if it is not repeated frequently.

Defining the Core Team. While firms pour thousands of dollars each year into improving processes for purchasing or order entry, and optimize these processes using process definition, Six Sigma, and Lean, most companies never define an innovation process, conducting innovation efforts in an ad hoc manner. While companies would never allow each business unit to define its own purchasing parameters and identify its own approved vendors and purchasing processes, they do allow each product group or line of business to adopt its own innovation methods and tools, and deploy people with little training and no centralized methods to create new products and services.

There is a better way. Just as purchasing is centralized to ensure that every acquisition in every product group or line of business is conducted in a similar and effective manner, a core innovation team can be defined to create and manage a common innovation method or process, while providing innovation capabilities and tools to anyone undertaking an innovation project. In this approach, a central team is responsible for managing and maintaining innovation methods, processes, tools, and capabilities and assisting product groups or lines of business when they need help on innovation tasks. A small core team can assist many different groups and ensure a more consistent, effective approach to innovation. Note that the core innovation team recommended isn't *responsible* for innovation, but for *defining* the common methods and processes for the organization to use.

Transition Points. Further, many innovation programs falter at important transition points within defined processes. Perhaps the most important transition point is between idea selection and product development. History is replete with examples of organizations that generated hundreds of great ideas that were never developed or implemented.

Xerox PARC is probably the best-known example of an organization that created many new innovations but failed to transition those concepts into new products or services. Xerox PARC is credited with prototyping the first computer mouse, the first graphical user interface, and a number of other technologies that were finally brought to market by other firms.[9] Xerox PARC struggled to move new ideas out of the research lab and into product development, and their struggles are reflected in many other organizations. Often the barrier is in the transition from idea selection to product or service development. The chasm between well-received idea and funded product development is large and it should be bridged by idea sponsorship, priority setting, and funding.

Ideas that are valuable to an innovation team and solve customers' needs may not receive the appropriate ranking or prioritization from an overworked product manager with a long list of priorities. This issue must be solved by integrating the product development team into the idea development, so the product managers understand the value and opportunity the idea presents. Also, sponsors need to be identified who can support, fund, and ensure achievement of the idea, placing correct prioritization on the product manager's to do list. It's not enough to document a process to generate and manage ideas, the process must consider key "gates" and decision points like funding and important "gaps" or chasms like the transition from idea to new product or service development.

Reactive versus Proactive Philosophy

While some of the factors we've considered (compensation, metrics, and processes, for example) are the outcome of intentional, careful decisions and specific actions, some of the factors influencing innovation are often derived over time or they are an artifact of the history of the organization, its position in the market, and its strategic focus. For example, many firms adopt the strategic position of "fast follower," discussed earlier, intending to enter new markets or create new products once those markets or product spaces have been validated by a competitor.

Far too frequently, many firms settle for such a "reactive" approach to innovation, using it as a tool to respond to changes in market conditions and in response to new entrants or new offerings, rather than using innovation in a proactive way to open new markets or address unmet opportunities. An example from one of OVO's clients is instructive here.

OVO Client Case Study. We worked with a large health insurer prior to the 2008 presidential election that brought President Obama to the White House. For at least two years prior to the election it was clear that a Democrat would win the White House and when he or she (Hillary Clinton was considered the frontrunner at that time) did win, one of the first big priorities would be to address health-care delivery and funding. While we raised this issue repeatedly, our client decided to take a "reactive" wait and see posture, rather than using innovative tools to create novel solutions that could become the leading thinking in the marketplace.

The argument given by our client was that regulations could change quickly and any new ideas or products they created would be worthless if the administration and Congress worked in directions different than the insurer chose to pursue. This attitude, which is prevalent in many firms, simply ignores the concept of trend spotting and scenario planning, and instead asserts that the future is basically unknowable. It is possible, however, to consider potential futures, and with a bit of work and insight any firm can predict with great confidence where markets, regulations, and customers are going. Good innovators don't react to the market, they identify emerging opportunities and arrive with products and solutions before customers are aware that their needs or conditions have changed.

To continue the saga of OVO's health insurance client, a year after the inauguration of President Obama, the insurer had fallen afoul of congressional Democrats, who sought to control increasing insurance prices. With no new products to introduce, our client had to resist new legislation and take its lumps in the media and in its markets. By ignoring the potential future and taking on a purely reactive mode, this firm lost several *years* of opportunity for innovation and now must work within the confines of the new legislation, reacting to regulations, rather than having acted in advance to influence the legislation by demonstrating innovative new products and services.

Firefighting. Most entrepreneurial and smaller firms want to change the world, and they are constantly trying to influence the dynamics of the market in significant ways. Over time, as the firms age and settle into a comfortable BAU existence, the expectations shift. As firms mature they seek to protect their markets and drive out costs and inefficiencies. Companies also become more defensive about their markets and prefer to *react* to changes rather than *create* changes. In fact, many firms in an industry try to codify the status quo, locking in existing rules and expectations and locking out new entrants. When the market inevitably shifts, most of these firms are caught off-guard, and resort to "firefighting." Firefighting is a term I use to describe the urgent demand and rapid response to a new product introduction by a competitor or a significant change in the marketplace. Middle managers are often asked to drop everything they are working on and respond to events in the marketplace.

In many firms it is hard to distinguish middle management from firefighters since middle managers are typically stuck with rushing from one fire to another. These fires are caused by unanticipated changes in the market through new regulations or entrants. Because the firm has settled on a reactive posture, and isn't actively attempting to influence the market or understand the future, every new change is a significant hurdle that must be addressed. Most firms reward "firefighters" who rush to handle these issues, though in hindsight these problems could have been avoided by a little *foresight* or proactive efforts.

A true innovator will identify these emerging issues and create new products or services to forestall change or to influence the change to favor their products or solution. Innovators are proactive, establishing new markets, identifying and meeting emerging needs well before the "fast followers" or laggards. Innovators force their competitors to become firefighters, which expend their energy and resources to stay abreast of the latest products, simply hoping to keep pace with the innovators.

> "Firefighting" seeks to bring the firm back to the status quo, whereas proactive investigation and innovation seek to advance the firm to an entirely new position.

Trend Spotting. Any firm can invest a small amount of money and resources into trend spotting. Trend spotting involves identifying changes that are occurring in technologies, economies, demographics, and other fields that will influence future markets. Trends may suggest that the demographic nature of a country is shifting, becoming older and more homogeneous; that economic growth is slowing; or that new technologies will dramatically change the way people interact with each other. Trends are easily spotted if people are alert to what happens in their markets and economies. Many Web sites, such as www.psfk.com or www.trendwatcher.com, collect and publish trends.

Trends are valuable as they indicate the potential direction and shape of the markets in the future. While trends don't necessarily indicate exactly what will happen, they provide clues as to the emerging opportunities and threats that may exist in your markets. Even if your firm doesn't care to do this work in house, it can easily be outsourced. Once you have the trends as inputs, conducting short scenario planning exercises to try to understand the future will begin to shift the firm into a more proactive stance.

Scenario Planning. Scenario planning isn't difficult and it is a great tool to begin to anticipate possible futures. Scenario planning uses the trends you've collected as inputs. Using those trends, your team discusses the impact of trends in a number of areas—technological, demographic, societal, governmental, and so on, and forms hypotheses about the future based on those trends. Thus, scenario planning helps your team create alternative views of the future. Forecasting in this way can help your firm predict market shifts, identify emerging market opportunities, and anticipate new entrants. You can learn more about scenario planning from perhaps one of the best sources; Peter Schwartz introduced scenario planning at Shell Oil in the late 1960s, and he eventually wrote one of the most approachable books on the subject entitled *The Art of the Long View*.[10] Once your team begins to understand the potential futures, then you can become proactive, influencing the market or regulators if necessary, creating new products and services that address the needs as they emerge.

From Reactive to Proactive. Shifting the focus for middle managers from "firefighting" to future scanning is possible and they'll appreciate the change. Firefighting is taxing, frustrating, and rarely leads to good outcomes, while requiring a huge expenditure of resources and resulting, at best, in a return to the status quo. Instead of moving the firm into a new or better position, firefighting is an investment that optimally returns the firm to its original status quo. Future scanning, however, shifts the focus toward the future and new opportunities, placing the organization in a position where it can be proactive.

Middle managers will again take their cues from what executives say is important, what key goals and metrics are communicated, and their personal evaluation and compensation plans. Executives who demand scenario planning to help shape the course of the business will introduce the importance of scenario planning and future scanning, and those tools will cascade throughout the business. Just as important, firms need to downplay the "heroism" of fighting fires and instead reward managers who spot problems or opportunities *before* they occur. Good innovators are proactive and they use their insights about the future to take advantage of the market and their competitors, moving into valuable positions ahead of other firms. Reactive companies constantly fight fires and long for the day they can steal a march on their innovative competitors.

Here, again, clear strategy and intent is important. A firm that lacks a well-defined strategy or has a strategy that isn't well communicated will shift its focus to what it knows best. In the absence of a new strategy, sustaining and improving the operating model *becomes* the strategy. The heightened focus on the operating model means that the firm cedes the proactive space to other firms that focus on innovation, trend spotting, and scenario planning, and it becomes ever more reactive to market shifts and consumer demands. The inevitable disruptions, when they do occur, create more dissonance for the reactive firms—which didn't foresee these changes and who are locked into a highly efficient operating model—than they do for proactive, innovative firms.

Human Resources and Talent Management

Finally, let's examine a component that is frequently overlooked, but it is certainly not the least important. We can call this factor a simple mnemonic—the three "Rs"—which consists of recruitment, retraining, and rewards and they are directly related to human resources and talent management. Most innovation teams pay lip service at best to these factors, rarely incorporating these insights or capabilities. Yet most of the cultural roadblocks and barriers for innovation are the responsibility of talent management and human resources.

Innovation relies on people more than other processes. This reliance on employees, management, and executives in an organization requires that the "right" people are attracted, and then given the appropriate tools and techniques for a sustained innovation success. Their passions and capabilities also must be ensured to align with the needs and expectations of the firm. Let's take a closer look at the three Rs.

Recruiting. The first way to transform a BAU culture to a more innovative one is to change the kind of people you hire. As a firm grows and matures and the "operating model" becomes widely accepted, executives and managers tend to hire people who have the same kinds of perspectives and experiences as those that exist within the firm. In that way training needs are reduced and the learning curve is shortened; however, the risk of "groupthink" increases dramatically.

Innovators understand that introducing people with deep skills but with different or complementary perspectives adds value to the existing "operating model" but it also adds new emphasis to innovation skills. These recruits may ruffle a few feathers, but they will introduce a significant number of new ideas and perspectives. Recruiting even a few "creatives" or right-brained thinkers into a rigid left-brain company can add just enough dissonance and creative tension to start shifting the thinking of the company as a whole.

This shift in thinking will occur as the creatives begin to question the status quo and the perspectives of middle managers and executives. Introducing people with different experiences and different perspectives will create dissonance and force the existing management structure to reconsider its objectives and perspectives, while introducing new concepts and ideas that haven't been considered within the firm previously.

Recruiting new people with different cultural backgrounds and skills is relatively easy. Finding appropriate homes where their skills will be accepted and rewarded in a more conservative corporate culture, however, may be more difficult. A culture that prides itself on rational thought and quantitative thinking will be tempted to squelch creative, right-brained thinking. Therefore, the recruiting activity must also be tied to finding appropriate roles and opportunities for the creatives within the organization to give them a chance to impact the culture. Even if a few new innovative or creative employees are hired, the vast majority of the firm remains. This remaining group will need to gain new skills, perspectives, and a focus on innovation. That's where the second "R"—retraining—comes into play.

Retraining. A program of training or "retraining" is valuable for tenured employees. This training is focused on fostering creativity and innovative thinking, and it will require a shift from much of the focus over the last decade, when efficiency and effectiveness have been the norm. That's why we use the "tongue in cheek" title of *re*training. While many managers have received a significant amount of training on tools designed to cut costs and optimize processes, they've received little or no training on innovation. Innovation tools and techniques will, in many cases, conflict with the optimization and efficiency training they've received in the past.

Introducing a number of creativity and innovation tools and techniques, alone, however, won't generate more innovation. New training and new perspectives are required to make people aware of their expectations, perspectives, and "anchors" that limit them from thinking creatively and innovatively. By "anchors" I mean the firm expectations, rules, and accepted wisdom that govern most of a middle manager's thinking and that limit the perspectives and ideas they are willing to consider. Until these perspectives and anchors are changed, or at least until we've made people aware of the barriers they impose on their own thinking, innovation and creativity remain difficult to accomplish. Talent management and human resources control much of the training budget in many companies and they are often responsible for defining appropriate training materials and curricula. Thus, training the existing executives, managers, and staff on the importance of innovation and on new tools and techniques for innovation can begin to shift the culture as well.

Innovation training can be delivered as a series of ongoing courses, or, perhaps more helpfully, in "just in time" offerings tailored to the needs of the innovation team. The training can embrace different aspects of innovation, from creative thinking techniques to deep investigations of specific innovation tools or methods. The range of options is vast. Hundreds of different packaged training programs exist on creativity, innovation processes, and tools and programs. What's important in this step is to decide the strategic goals and core competencies of the firm, and then select the innovation training programs that will help achieve your goals.

Rewarding. The final impact talent management and human resources can have on innovation is in reward structures. There are several factors to consider in reward and compensation programs. First, consider the "regular" compensation programs that focus on the standard compensation schemes. Too often compensation is tied to expectations of efficiency and effectiveness, so not only compensation, but evaluations must be modified as well. Other kinds of rewards and recognition must be considered as well. Research shows that many innovators find intangible rewards for innovation equally as compelling as financial rewards, so structuring a broad rewards program for innovation is vital.

Intangible reward systems can include a wide range of actions, including

- Simple recognition systems, such as identifying an "Innovator of the Year"
- Allowing people to work on ideas that align with their passions
- Offering new roles or titles aligned to innovation
- Offering new training opportunities to people who excel at innovation

Much enthusiasm can be generated for innovation with little financial investment if rewards other than standard compensation are carefully developed.

> Ultimately, innovation is successful or unsuccessful based on culture, communication, skills development, evaluation, and rewards, all of which are influenced by the individuals least likely to be part of an innovation project: HR and talent management.

Get the HR Team Involved. Tap your talent management and human resources teams to improve your team's skills, and to encourage more innovation by changing the recruiting profiles, training options, and evaluation and compensation schemes. Human resources has traditionally been viewed as an overhead cost rather than a strategic contributor to the business. Where innovation is concerned, human resources and talent management must have an equal seat at the table, because so much of the culture and "operating model" of an organization is directly or indirectly influenced by human resources and talent management.

Innovation BAU

Throughout this book so far I've argued that business as usual stymies innovation by refocusing the organization on efficient, effective business processes and by commanding the attention of the middle management cohort. In this chapter I've examined a number of attributes that you can change to introduce far more innovation into your BAU culture and methods, with the ultimate goal of creating a new business-as-usual culture—an innovation business-as-usual focus.

The goal of this new BAU is to bring the operating model back into balance, focused on both efficiency and innovation. As I've described, the shift toward efficiency and effectiveness has been ongoing for close to two decades, so the shift back toward a more balanced model will take time and focus. The eight factors I've described in this chapter are all important in the effort to rebalance the model, and they must be reinforced by the people who are responsible for keeping the BAU operating at peak efficiency—the middle managers. In the next chapter we'll examine how to help middle managers shift their focus and their skills from efficiency to a balance between efficiency and innovation.

Chapter 8

What Happens to the Middle Managers?

Sometimes it seems that middle managers receive all the blame for the failings of an organization, while never receiving accolades for the many successes they create. Middle managers have traditionally defended the organization from large threats such as market disruptions and new entrants, and small ones such as unhappy customers or inefficient processes. They manage large organizations with little authority and lots of indirect power, aided by robust corporate cultures that seek to keep people and ideas in line. Middle managers translate the big picture visions and ideas of executives and create action plans for front line workers, filtering the feedback from customers and front line employees to executive management. Middle managers are present in every decision, communication stream, funding cycle, and customer and partner interaction. Little work of any significance happens in a large organization without the involvement of MM. Likewise, no significant change is created or implemented without the complete acquiescence and buy in of middle management, which is why the role remains a powerful barrier to innovation.

In previous chapters, I've defined the importance of BAU and how middle managers cultivate it and use it to their advantage to keep a vast number of products, initiatives, decisions, and people working as efficiently as possible. In the last chapter I introduced the concept of an innovation BAU in which innovation becomes a persistent capability and discipline. In this chapter we'll examine the approach necessary to shift the thinking of middle managers from resisting innovation to fully supporting innovation, based on the shift of BAU to innovation BAU.

How Middle Managers Become Supportive of Innovation

The idea that middle managers can be innovation champions is not a new one. Rosabeth Moss Kanter published *The Change Masters: Innovation and Entrepreneurship in the American Corporation* in 1983, which focused on the importance of middle managers as engines of innovation. Over 30 years ago Kanter documented the important role that middle managers play in innovation efforts. She found that middle managers who support innovation share five characteristics:

- Comfort with change and uncertainty
- Foresight, recognizing unmet needs as opportunities
- Thoroughness, well prepared with a good understanding of organizational politics
- A participative management style
- Endowed with persuasiveness, persistence, and discretion

Further, the innovative middle managers she identified understood the value of time and tact to accomplish their goals. They were made more successful by an organizational culture and structure that fostered innovation.[1]

More recently, in *The Innovator's DNA*, published in 2010, authors Dyer, Gregersen, and Christensen identify five key skills that innovation leaders demonstrate:

- Associating
- Questioning
- Observing
- Experimenting
- Networking[2]

Interestingly, I'll argue that many of these skills are necessary for efficiency and effectiveness *as well as* innovation, but the perspectives to meet these standards are much different. Take, for example, experimenting. Managers often conduct small, focused experiments to improve efficiency. Where innovation is concerned, those experiments need to become something more like explorations: opportunities to discover new insights, rather than simply validate what's known.

For MM, perhaps the most vital and most neglected skill of those identified in *The Innovator's DNA* is networking. More research from Martin Ruef and Richard Burt demonstrate that managers with "diverse, horizontal social networks that extend outside their organization and involved people from other diverse fields of expertise" were three times more innovative than managers with "uniform, vertical networks."[3] Yet far too often middle managers focus on solving internal problems using internal resources, rarely building broad, horizontal networks, or incorporating insights or knowledge from people in other companies or industries than their own.

If we constantly remind ourselves that MM's first priorities are the short-term goals of the business, then it becomes apparent that middle managers will strive to innovate only when it is perfectly clear that innovation is at least as important as efficiency and short-term profitability. The first logical conclusion we can draw from this scenario is based on one of the attributes of the culture we described in the previous chapter: *Middle managers focus on strategic imperatives that are well communicated, carefully planned, clearly important to the business, and consistently measured.* If innovation is introduced and reinforced by these factors, middle managers must contend with two competing but equally important goals: efficiency and innovation.

Middle Managers' Next Steps

Once the realization sets in that innovation is equally as important as efficiency, middle managers will take the next appropriate step, demanding the best tools, techniques, and training for themselves and their teams in order to be as proficient as possible at generating and managing ideas. Middle managers are compensated for achieving goals and using resources effectively. If innovation is an important, measurable goal, they'll want to find the best, most efficient ways to conduct innovation efforts, leading to one of several outcomes. Besides demanding training to build skills and capabilities for themselves and their staff, MM may seek out experienced consultants who can develop innovative ideas quickly and competently.

Middle managers will recognize their own shortcomings and those of their teams where innovation is concerned. In many cases they may seek to find experts who can deliver innovation capabilities and insights quickly. As discussed, working with consultants, while reducing risks and shortening timeframes, introduces significantly higher costs and provides little knowledge transfer. I've worked with one client for several years who refers to this phenomenon as "buying ideas from consultants." While there's little knowledge transfer in this approach, many middle managers may turn first to consultants to speed innovation efforts initially. Innovation consultants can reduce the impact on the existing business processes and create more radical ideas. Alternatively, there is a precedent for middle managers to receive training to deploy a new capability.

Over the last decade many middle managers have attended classes and learned the methodologies and techniques that sustain Six Sigma and Lean. Today there are as many "black belts" in many organizations as there are in a Bruce Lee movie. If the time and resources can be found to train managers on Six Sigma, Lean, and other efficiency tools, certainly the same resources can be found to train middle managers on innovation tools, methods, and skills. While this training and the implementation of the methods and processes may be time-consuming, it is a reusable capability that increases its value the more frequently it is used. Just as internal teams have gained knowledge and increased throughput and efficiency over time in traditional processes, internal teams that are adequately trained and fairly compensated can create interesting and radical ideas with little input from external parties.

> In many firms there are as many black belts as there are in a Bruce Lee movie. Can we develop a corresponding number of *innovation* black belts?

Middle managers can also choose to pursue "open innovation" and work with internal teams or external customers and partners to generate, rank, and test ideas. Open innovation doesn't eliminate the need for well-understood and well-defined innovation processes within a business, but it can widen the scope and range of ideas that are generated and assist with market validation, prototyping, and other innovation-related tasks. With an engaged customer base, open innovation can create many ideas in a short period of time, and provide some sense of the enthusiasm behind the ideas, reducing the evaluation time and risks of choosing the "wrong" idea.

Whether the innovation activity is completely insourced or innovation activities are conducted with consultants or through open innovation initiatives, middle managers must ensure their teams and processes can support and sustain ideas through the entire innovation cycle. This means that regardless of the origin of the ideas, internal teams need skills and capabilities to evaluate, assess, and develop the ideas as new products or services. Internal skill development is important irrespective of the method of idea generation selected.

Finding and Creating "Multidimensional" Middle Managers

The challenge facing any organization that seeks consistent innovation is to shift the "operating model" to balance efficiency and innovation. To accomplish that shift, you'll need to shift the capabilities and the focus of middle managers who need to embrace both efficiency and innovation, becoming multidimensional. Let's look at how your firm can identify, attract, and retain more multidimensional middle managers who will drive innovation success.

> Assess existing middle managers and discover their aptitude for innovation. Recruit new managers who have innovation experience or strengths.

Assessment

Start by understanding the skills and capabilities of your existing middle managers by assessing their interests, capabilities, and skills. Many firms assess their employees with templates like Myers-Briggs or the DISC model. Why not assess your managers based on innovation capabilities and preferences?

Start this effort by identifying managers in your organization who have innovation proclivities and skills. This can be done with something as simple as a set of assessment questions about interest and openness to innovation. If your team desires more rigorous quantitative assessment tools, there are several formal programs to choose from. One of those assessments is the "KAI" assessment, the Kirton Adaption Innovation Inventory. The Adaption Innovation research is based on psychometric profiling to discover how people prefer to solve problems and their ability to apply creativity in doing so. The KAI Index has been used successfully by many firms to assess the innovation and creativity potential of their employees.

Another assessment I've used with great success is the Foursight assessment, developed by Gerard Puccio, Ph.D., who is the director of the International Center for Studies in Creativity at the State University of New York College at Buffalo. The Foursight assessment suggests that every employee can play a vital role in innovation, but their interests and skills may support different phases or needs of the innovation process. The assessment places an individual into at least one of four types—clarifier, ideator, developer, and implementor. With this assessment in hand, your team can identify the individuals who are best at specific tasks, while ensuring the innovation initiative is populated with a diversity of skills and perspectives.

The Innovator's DNA provides another type of assessment, examining key skills and attributes that innovators share. The book and accompanying assessment can also help to identify the best innovators in your organization, and help spot innovators in your application pool.

While everyone can and should participate in innovation, some individuals have more interest, proclivity, and capability for innovation than others. Identifying the middle managers who prefer to innovate and using those individuals as the pioneers of your innovation effort will help accelerate a successful shift in your operating model.

How Many MM Innovators Do You Need?

The next question your firm should explore is: how many "innovative" or "multidimensional" managers are needed for a successful innovation initiative? Does every manager need to be good at innovation, or can you be successful if only a subset is trained and capable of supporting innovation? That question is dependent on the commitment of your organization and the nature of competition in your industry. Ideally, every manager would be able to demonstrate at least *some* innovation capability and competency, but that may take years to achieve.

Assess your market and industry. If the pace of change is accelerating, and many new entrants are attacking the market, you'll need far more innovation capabilities than you have now. If your market is a bit more stagnant with few changes, you'll still want to have a number of managers who are focused on innovation, but perhaps your organization can move more slowly with its efforts and with building up your innovative MM.

Further, your firm will need to decide how to introduce innovation capabilities and skills and how quickly those skills are needed and can be learned. In addition, there are a range of innovation skills that are necessary in any innovation effort. The skills range from trend spotting and scenario planning to ethnographic and research skills to facilitating idea generation sessions. Few middle managers will be proficient at all of these skills, so it makes sense to "inventory" the skills of your middle managers, so you'll know who to call on when you need specific competencies.

Back to the Three Rs

Once you've identified the skills and capabilities of your core middle management team, and determined the depth and breadth of innovation skill your firm needs, turn your attention to the three "Rs" discussed previously: recruiting, retraining, and rewarding.

Clearly, you can recruit new people to your organization who have more innovation skills to augment your predominantly efficiency-oriented management team, but you'll need to define new roles and responsibilities to attract and retain those new employees. Those roles need to be meaningful and valuable for the individual and for the business. Be careful, though: placing a manager who has a bias toward innovation in a role that demands and measures effectiveness and efficiency will frustrate both the organization and the individual. Creating a "make-work" job to retain an individual with deep innovation skills will not achieve value for either party.

Consider new roles, for example, an "ideas" manager, as a corollary to product managers. Today, there are two well-defined manager roles: a product development role and a product manager role. The development role has the responsibility to translate requirements into a new product or service. Once the product is commercialized, a product manager markets the product, gathers new requirements, and develops a roadmap for the product. I'll argue that one critical role is missing—the role responsible for identifying and managing ideas until they are ready for development.

While the need is evident, few firms have identified an "ideas" manager, who spots opportunities and develops a range of ideas. The ideas manager could "own" the front end of the innovation cycle, handing off ideas to product development, which then hands ideas to product management. This "ideas manager" is simply one example of a role that may attract and retain more innovative middle managers to your organization and help balance the skills and capabilities of the MM team.

For innovation success, consider training or retraining your existing middle managers to balance their efficiency and effectiveness skills with innovation tools, techniques, and methodologies. The costs of training are in two parts: the financial outlay to pay for the training, whether it is self-paced, in-house, or external; and the cost of the time away from managing the existing process. Neither of these costs by themselves is steep, but combined the return often doesn't seem worth the investment. However, the investment in training sends this clear message to the middle managers: *innovation is as important as efficiency, and we can't afford ineffective innovation.*

Note that the range of innovation skills and capabilities is broad. Not every manager should be trained in every tool or technique. It is important for your firm to identify the innovation tools, methods, and processes that are central to success within your organization and train broadly on those. Then, you must allow individuals who have interests or proclivities for more specialized techniques like scenario planning, TRIZ, or needs-based innovation to pursue those through external programs and reference books. A core innovation team that acts as a repository for innovation processes and tools can help define a small number of key innovation methods that your team should master, and it can be a resource for other tools and techniques that your team needs only occasionally.

Finally, consider how your firm evaluates, compensates, and recognizes its middle managers. Most organizations have fine-grained compensation schemes developed by human resources that restrict compensation by grading, bands, and evaluations. In order to encourage more innovation in your middle managers, just as throughout the rest of the organization, innovation must be measured and managed in the evaluation cycle, and the efforts associated with innovation must be recognized in the compensation program. As many innovation experts will attest, good innovators thrive not only on financial compensation but also on recognition, rewards, and the ability to work on their best ideas. The compensation schemes must therefore also include recognizing the best innovators publicly and, where possible, allow them to participate in the development of their ideas.

The Multidimensional Manager

Once you've assessed your team, decided how to train your middle managers in innovation tools and techniques, and changed how they are compensated and evaluated, you are on your way to creating more "multidimensional" managers. But assessment and training aren't enough—these managers need to implement their new skills and capabilities or their previous reactive perspectives, protecting the status quo, will rapidly creep back in. When middle managers quickly deploy their new innovation skills, their perspectives begin to change. They seek new opportunities for innovation, and they begin to shift the operating model to incorporate more innovation focus.

As this shift occurs, it is critical that the executive team's communications continue to balance the importance of innovation and efficiency. Likewise, the compensation programs, evaluation systems, and other assessments must demonstrate an ongoing balance between innovation and efficiency. The first few weeks and months of the transition are critical. The shift away from a consistent, comfortable model is just beginning, and slipping back into the well-known operating model is quick and easy. Any signal that innovation isn't as important as efficiency and short-term goals received by the middle managers will cause them to revert to previous management styles, and resist innovation.

Only with constant reinforcement will the model shift permanently into a balance between efficiency and innovation, and middle managers will dictate whether or not that shift occurs. In this regard, new innovation-oriented middle managers are hot-house flowers, requiring a consistent environment to take root before being exposed to the vicissitudes of management and the market. Only as the middle managers have time to develop their long-neglected innovation skills and balance those skills with well-developed efficiency skills will the BAU model shift.

Engaging with All Deliberate Speed

Developing the skills and perspectives necessary to shift an organization to a focus balanced between innovation and efficiency must be planned and executed in a timeframe that allows middle managers to come up to speed on new tools, methods, and perspectives. Too often, important innovation goals aren't achieved because the shifts are introduced far too quickly, before the middle managers have a chance to "come up to speed." The individuals who need to lead the effort, and their teams, simply don't have the time to gain the skills necessary to innovate successfully. Let's look at what it takes to engage the shift from a focus on efficiency to a more balanced focus, once again using a sports analogy. In this instance consider the middle manager as an athlete.

Over the last 20 years middle managers have trained to become organizational "sprinters," constantly running the same race each quarter, a sprint to produce consistent quarterly results that meet or exceed expectations on Wall Street. The focus on efficiency and achieving financial goals quarter on quarter means that their only real athletic skill is short sprints. When organizations place emphasis on innovation, they ask these athletes, who are excellent sprinters, to compete in completely new events, with little training or preparation. Innovation, however, should not be considered a "sprint," but a decathlon.

The Olympic decathlon is meant to identify athletes that excel in a wide range of track and field events—sprinting, jumping, hurdling, throwing, and distance running. Only an exceptionally well-trained athlete who has this array of skills can compete effectively in the decathlon. Just like such a decathlon, innovation requires rapid starts, like a sprint, along with stamina and discipline, like a distance event. New innovation efforts will require "hurdling" obstacles and throwing out old approaches or perspectives.

Imagine that dedicated middle managers, who have spent years perfecting their short sprints, are now called on to compete in a completely different event—one that demands skills, techniques, and talents they may possess but certainly haven't practiced recently. They may also have had little to no coaching or support from their organization.

In the decathlon, an athlete who is only prepared for the sprints can effectively compete in *one of ten* events. In the other nine events, the sprinter must simply do the best he or she can. Likewise, middle managers thrust into an innovation role must adopt innovation tools and techniques as quickly as possible, and they must do the best they can with limited training and little to no preparation. We'd never compete like this in the Olympics, yet we ask our middle managers to "compete" in fiercely competitive markets with little innovation aptitude or training. Clearly there's a better way.

Depending on several factors, including the amount and distribution of training, the size of the organization, and the communication of the intent and goals, the transition period to engrained innovation could take anywhere from several months to more than a year to complete. Without this transition period, allowing the middle managers and their teams to gain new skills and try out those skills in controlled settings (practice), middle managers will be called on to achieve innovation goals for which they don't have the skills, perspectives, or inclination to achieve. Establishing a reasonable transition period to allow the skills and perspectives of middle managers and their staffs to shift enables these teams to conduct interesting innovation work effectively, and shift the operating model into balance. Attempting innovation without the transition period is difficult and often results in outcomes that are incremental at best.

The Shift for Middle Management

Starting the shift to innovation BAU must begin with changing the vision, perspectives, skills, and goals of middle managers. As innovation is introduced strategically, middle managers must receive training on new tools and methods, and begin to deploy the training in carefully supervised initiatives. Only as the strategies, tools, and perspectives are introduced and middle managers have the chance to gain skills and competence, should the organization scale up innovation efforts and demand large-scale innovation projects or disruptive innovation.

In the future, as innovation becomes a more consistent and persistent business process and discipline, middle managers will need to establish the ground rules and constantly reinforce the environment necessary for innovation to flourish. They will have to be familiar with important innovation tools and techniques, and document the innovation methods and processes to speed ideas through the innovation funnel—all while achieving important quarterly financial goals. This combined need for innovation and for achieving the short-term goals, will require existing middle managers to adopt new ways of thinking—to become ambidextrous—and it will require new workers entering middle management to embrace innovation from the start of their careers.

As innovation becomes business as usual, the role of middle managers will shift again. Increasingly they become responsible for maintaining a balance between efficiency and innovation in the operating model. Further, they are responsible for developing the culture in which innovation can thrive and developing the skills and capabilities of the staff workers who will generate and manage ideas.

Chapter 9

Shifting Your Operating Model to Balance Efficiency and Innovation

Throughout this book, I've presented Google, Apple, 3M, Procter & Gamble, and Gore as successful, relentless innovators. What's surprising is how different these firms are from one another. They are in different markets, with unique characteristics and offerings. Some have charismatic leaders and others have CEOs you wouldn't notice walking down the street. Some participate in highly competitive markets with aggressive competitors and others have a large share of their market space.

These innovation leaders are dramatically different in how they derive sales and profits:

- Google's from advertising
- Apple's from selling consumer electronics and content
- 3M's from basic consumer and industrial products
- P&G's from consumer products
- Gore's from technologies

One could argue the only visible factor that connects these firms is their ability to innovate successfully over time.

Yet if you drill a little deeper into their success you'll see some startling similarities, primarily around the expectations their cultures have about innovation, and how the middle managers and front line employees participate in innovation activities. Consider, for example, how Google embeds innovation in its culture and processes.

Google's Innovation BAU Culture

Anyone at Google can create and develop ideas. The company borrowed the concept of "20 percent time" from 3M, which allows Google employees to spend 20 percent of their time working on interesting ideas.[1] Most Google and 3M employees will tell you that the time is often spent on their own time (outside of regular business hours), but both companies embrace individuals and small teams that create valuable new ideas by providing time and resources to help develop those ideas.

At Google, a good idea is identified by how many other people you can recruit to help you develop your idea. In a company in which everyone is a potential innovator, Google uses an interesting investment model to help decide which ideas are valuable. If everyone has 20 percent of their time available to apply to any idea, the employees become highly selective as to where to invest their time. The best ideas end up attracting the best people, who willingly commit their time to the potential new product or service. Google also fosters a "sense of fearlessness"[2] in its employees, encouraging them to create new ideas. The company's rewards systems also lead to innovation. Some Google engineers who had great ideas received large stock option grants when their ideas developed into actual products or services. This award is called the Google Founder's award and it has been worth as much as $12 million.[3] While encouraging and sustaining consistent innovation, Google's BAU processes are still meant to drive profitability and revenue.

Considering the list of factors for innovation success, it is possible to see that Google meets these criteria:

- Google executives have clearly defined goals
- The culture actively encourages innovation
- Everyone at the company understands the importance and purpose of innovation
- Compensation structures reward innovation
- The culture itself encourages fearlessness and risk taking

Google doesn't have a magic formula for innovation; it has an innovation BAU culture in which middle management and front line workers are encouraged to innovate while consistently achieving quarterly results. Procter & Gamble is another firm that demonstrates consistent innovation over time, yet differs from Google in many ways.

Procter & Gamble's Consistent Innovation

Procter & Gamble is a much older firm than Google, in a different industry segment, and it manufactures its products and services on a global basis. Yet P&G exhibits many of the same outcomes as Google. For example, the firm has a clear management position on innovation. While Google has traditionally generated and developed many of its ideas on its own, P&G actively encourages partners and customers to work with its internal teams to generate new products and services. Yet, when you drill into P&G's teams and processes, you'll see many of the same factors at work as in Google.

CEO A. G. Lafley established a stated, measurable public goal for innovation at P&G, and other innovators, such as Chris Thoen, a previous managing director of the Global Open Innovation Office at P&G, carried on that focus. While Procter & Gamble relies much more heavily on open innovation than Google does, the internal processes and capabilities to develop and manage ideas exist within Procter & Gamble just as they do within Google. Clearly, in both Procter & Gamble and Google there is a cadre of middle managers supporting innovation capabilities and processes. Procter & Gamble's middle managers may focus more on technology scouting, business development, partnering, and open innovation than their Google counterparts, but managers from both firms are engaged in innovation activities, which are supported and enabled by the culture, reward systems, and firms' expectations. Though Procter & Gamble product managers must achieve quarterly results and compete in a market that is far more competitive than Google's, they continue to innovate. Let's take a close look at another innovator mentioned throughout the book—W. L. Gore.

W. L. Gore's Sustained Innovation

W. L. Gore is a technology innovator as well as an innovator of corporate structure. Gore has a very flat organizational structure focused on teams, rather than the traditional hierarchical structure. Gore's innovative structure and culture were investigated and analyzed by Gary Hamel in *The Future of Management.* He suggests that Gore represents the next evolution of corporate hierarchy, making Gore a management innovator as well as a technology innovator. The company encourages innovation based on its core technology, which has led it into markets and solutions as diverse as raincoats, dental floss, guitar strings, and vein replacements. Gore's philosophy encourages creativity, personal initiative, and open communications, so the firm acts much more like a start up than a multi-billion-dollar firm.

At a Front End of Innovation Conference in May 2011, Terri Kelly, the CEO of W. L. Gore, gave a presentation on the factors that sustain innovation at Gore. Perhaps the most important aspects she highlighted were the challenges:

- Balancing focus with entrepreneurial diversity
- Valuing innovation and organizational effectiveness
- Developing sufficient capability and capacity of leaders

Note the "AND" focus—innovation *and* efficiency. Further, she left the crowd with some critical insights:

- Constant leadership vigilance is necessary.
- Make sure all of your practices and policies reinforce an innovative environment.
- Don't let discipline be viewed as a constraint to innovation.
- Organizational choices are important (ask: what are you going to be good at?).
- Ensure balance of skills and support in teams.
- Embed innovation philosophy, tools, and disciplines throughout the organization.
- Handling failures sets precedent for risk-taking.[4]

All these factors play directly into the company's innovation efforts as they focus on innovation and efficiency—a concept that all successful innovators must embrace.

Finding the Perfect Balance

All the innovative firms discussed run efficient, effective internal processes, regularly achieving expected quarterly results while creating valuable new ideas. They demonstrate that the concept of balancing an internal operating model between the conflicting forces of efficiency and innovation is not only possible, but demonstrable. What does it take to balance innovation and efficiency so effectively? There are several key factors.

1. A Top-Down Focus on Innovation

The first factor in balancing innovation and efficiency is *a strong focus on innovation at the top, reinforced by stated goals that are consistently measured.* Any firm under the discipline of the financial markets will strive to cut costs, which allow it to achieve expected quarterly results. All of the innovators identified earlier aren't just innovation leaders, they also use capital and other resources at least as efficiently as their close competitors, if not more so. The emphasis on innovation therefore must be at least as large as the emphasis on efficiency and effectiveness, and that emphasis must be consistently reinforced, through strategy, communication, executive action, and cultural and operating model change.

In the absence of these factors, the operating model, which was built for efficiency, will revert to its previous state, moving away from innovation and creating an imbalance. The forces that pull the model toward innovation must be more alert and more persistent than the inertia that will tug the model back into its comfort zone of efficiency. Further, the consistent participation from the top should include more than occasional communications. Examples from GE and Apple cited previously demonstrate that the CEOs of these firms are actively involved in innovation efforts and they constantly seek more information and insight about how their firms can innovate. Innovation starts at the top, but the "top" should stay involved and engaged in innovation beyond goal setting, communications, and measurements.

2. Embrace Tools, Techniques, and Methods

The middle managers and innovation advocates in an organization must embrace innovation tools, techniques, and methods broadly. Trying to create innovation depth in one business unit or one product line while the rest of the organization remains focused on efficiency is exceptionally difficult because the teams have competing goals. Perhaps even worse is locking the innovation team into only one method or approach for innovation, or only one "fence" or horizon. Many organizations assume because they've established an "open" innovation program that allows customers to submit ideas, they've also provided a complete capability for innovation success.

The gaps in this thinking are quickly exposed when customers don't receive feedback, ideas don't become new products with any regularity, and internal innovators feel slighted or left out of the innovation process. At Google, 3M, Procter

& Gamble, and Gore, every employee, manager, and executive is a potential innovator and everyone leverages the innovative side of the "operating model" as well as the efficiency side. If only a select few people are innovators, or if only a few tools or techniques are deployed, the effort becomes unbalanced and ultimately unsuccessful. Can your company deploy as many innovation advocates as you have Six Sigma black belts?

3. Consider Innovation a Revenue Opportunity

You need to *consider innovation as a revenue opportunity, not as a cost structure.* Innovation won't be successful until the team understands that the ultimate goal of the effort is to create revenue, growth, and opportunity. The existing operating model is focused on cost reduction and efficiency. If your innovation goals are too small or too focused around costs and efficiency, innovation will be superfluous to the existing operating model and folded back in. Establishing clearly delineated goals for innovation, and measuring and reporting the outcomes, creates a value proposition for innovation and again balances the operating model between efficiency and innovation. As the operating model is shifted from almost a 100 percent efficiency focus, it needs to move toward some new focus or capability.

If innovation doesn't drive growth, differentiation, and profits, it simply duplicates the existing model, which will revert back to 100 percent efficiency. Further, "cost centers" that don't drive revenue are easy programs to cut. If innovation is viewed as a cost center rather than a revenue center, it is easy to trim funding from the innovation program, not only because it will be a new program, but because innovation is so often thought of as overhead rather than a revenue generating possibility.

4. Becoming More "Plastic"

Your firm will need to *become more open, flexible, and nimble*, what brain scientists like to call "plastic." Good innovators understand that exchanging ideas and perspectives with people in other "stovepipes" internally and in other external organizations creates more innovation capability. The more siloed your lines of business and product groups are, internally and externally, the more difficult it is to innovate. Good innovators understand that new ideas spring from diverse experiences. If your goal is to balance efficiency and innovation, you'll need to open up to more internal and external influences and establish broader "horizontal" networks. Further, your team members will need to become far more flexible and adaptable.

Over time people have come to believe that deep expertise in a subject area is important, resulting in many people remaining in the same job for years. Their expertise can become as much a curse as a blessing as they lose the ability to think beyond their skill set and fail to acknowledge market shifts that may void or threaten their specific knowledge. As the rate of change increases in markets and emerging economies compete on equal footings with advanced economies, the breadth and depth of knowledge and expertise expands exponentially. Your company's experts now compete with experts everywhere, simultaneously, all of whom are generating new insights and ideas.

Increasingly, remaining competitive and innovative is much more about anticipating change and adapting to it, rather than hunkering down and relying on deep expertise, which may be made obsolete more quickly than ever before. People who have a wider array of experiences, jobs, and connections are often more innovative and more adaptable than those who have, or have had, fewer. To balance efficiency and innovation, middle managers must be encouraged to gain new perspectives, increase their relationships internally and externally, obtain new skills, and perhaps take on new roles in their company.

5. Adopt a Rapid Experimentation Methodology

Your company *must adopt a "rapid" experimentation methodology.* Good innovators understand that rapid prototyping and experimentation will drive the development of new ideas more quickly than almost any other approach. In many firms today, experiments are carefully designed and developed over a long period of time to validate a specific hunch or theory. This experimentation works well if the theory asserted is proven successful. Otherwise the experiments are virtually useless.

You must shift your thinking to recognize that experimentation and prototyping is as much about discovery and new insights as it is about validation of internal perspectives and theories. Your firm must make it far easier to test ideas, gain new insights, and "fail forward." In the time it takes to plan, conduct, and assess the results of a carefully designed experiment, your team can develop five or six iterative prototypes, expose those prototypes to clients and prospects, and gain incredible feedback on the design and demand of the product. In this way, customers take on a greater role in the product definition and development process, and your team opens up to more perspectives and ideas. (Peter Sims addresses the concept of experimentation and rapid prototyping exceptionally well in his book *Little Bets: How Breakthrough Ideas Emerge from Small Discoveries.*[5])

6. Patience Is Necessary

Throughout this entire process, you'll *need some patience.* Creating this shift takes time, just as any significant change in a large organization will. Core operating models resist change by their very nature, and introducing new ways of thinking, tools, and strategies will take time to filter down. Inertia and resistance, while subtle, are far stronger than your management team may believe.

Lafley, for example, publicly announced his goal in 2000, after a year of internal discussion and preparation, and only achieved his stated goals six years later. His objective was a major change and completely transformed Procter & Gamble from an internally-driven R&D organization to an "open innovation" leader—your firm's shifts may not take as long, but you must be prepared for resistance and slow initial change that is only accelerated by early wins. The late majority will come on board eventually, but it may take several years before the preponderance of your organization is willing to effectively balance efficiency and innovation. Consider this process a revolution played out in stages.

The first stage is to recognize the need for balance between efficiency and innovation. Once the acknowledgement is made, the senior executive team must prepare the appropriate communications and strategic vision to direct the thinking of the team. Human Resources and Talent managers need to develop new evaluations, compensation, and reward structures to encourage managers and staff to get engaged in innovation activities. Senior executives and managers must identify key opportunities, emerging threats, and needs where innovation can develop new product and service ideas. Only after this work is done will the operating model begin to shift noticeably, and even after all of this effort the model will easily revert back to its original comfort zone unless these actions are consistently reinforced.

7. The Cortes Moment

Finally, you *may want to plan for a "Cortes" moment* to initiate the shift in emphasis. Most organizations have spent the last decade or more becoming comfortable with a focus on efficiency, effectiveness, outsourcing, and right-sizing, reinforced by evaluation and compensation schemes, with the added benefit of inertia and fear of change. This means that the shift in emphasis must be dramatic, widely communicated, and leave no other options. Hernando Cortes understood this concept.

When Cortes landed in the New World, he scuttled his ships to leave his conquistadors no option but to move ahead. There was no turning back to the older, safer locations or approaches. Your team, too, needs a Cortes moment when you burn the bridges to safe solutions and point out the new direction. Much like Lafley did when he announced the open innovation goals for P&G, your Cortes moment needs to broadly indicate the shift, point out the targets, and demonstrate management commitment to the approach. Otherwise, inertia, fear of change, and the desire to return to comfortable, familiar tools and processes will cause the team to revert to known efficiency methods and tools.

An Unfolding Case Study: General Electric

There's probably no better case study[6] to examine in the context of innovation than General Electric, which I'll argue is in the midst of the transition outlined in this book. Jack Welch crafted in GE an engine of growth through acquisition and established an overwhelming internal focus on efficiency. Welch demanded that each business be a leader in its industry, and promoted the idea of ranking employees, decreeing that the bottom 15 percent of employees be let go each year. Under his watch GE gained market share and its share price increased dramatically.

In late 2000, Jeffrey Immelt took the reins from Jack Welch and began a campaign to open up GE to external influences. Further, Immelt wanted to create a more engaged workforce and softened the edges of Welch's ranking strategies for businesses and people. After the attacks on September 11, 2001, GE's business, along with the entire market, plummeted. According to a paper by Harvard business school professor Christopher Barlett, Immelt pursued many of the strategies we've defined here, cutting costs and implementing Six Sigma and Lean projects throughout the business. According to Barlett, GE completed 6,000 Six Sigma projects with its health-care providers in 2002.[7] In 2003, Immelt recognized the markets were accelerating and GE needed to grow more quickly. He promoted Beth Comstock as chief marketing officer and began developing plans for more innovation strategy. In late 2003, Immelt and Comstock announced the Imagination Breakthrough program, meant to encourage GE leaders to develop ideas that could generate $100 million in new revenue within three years. Within one year of launching the Imagination Breakthrough program, more than 80 opportunities had been identified and by 2005, 25 had generated revenue. Imagination Breakthrough projects in GE Healthcare, GE Locomotives, and other GE business lines have achieved or surpassed the $100 million mark. To date, several of these Imagination Breakthroughs have achieved the goal of generating $100 million in revenue in three years or less.

While Immelt may have been persuasive about his vision and the Imagination Breakthrough goals, the results weren't immediately satisfying. When he launched the Imagination Breakthrough program he told the heads of the business units invited to the event "I want game changers. Take a big swing" and gave them two months to get started. Quoting here from an article in *BusinessWeek*: "Some people in the room were stunned. 'There was a collective gulp across the organization,' Comstock recalls. 'People were thinking, "Is this real?"'"[8]

What was it about GE that caused the executives to react in such a way to Immelt's message? "Immelt worried that GE's famous obsession with bottom-line results—and tendency to get rid of those who don't meet them—(would) make some execs shy away from taking risks that could revolutionize the company."[9]

Since Immelt became CEO, he and the GE team have done a number of things right to foster innovation. Immelt established a clear organic growth goal (nearly double what it was under Welch) and demands innovation from the individuals who lead business units. As CEO he regularly reviews the top 35 Imagination Breakthrough projects, demonstrating his commitment to innovation. Along with his marketing team, he constantly reinforces the need for more innovation internally and positions GE externally as an innovation leader. With his talent management team, he actively improves the capabilities and knowledge of his organization through training oriented toward building innovation skills and through new hiring of people with fresh innovation skills and different perspectives. GE has also changed how people are evaluated, implementing compensation models and incentives to encourage more innovation. Immelt was quoted as saying about business leaders within GE "you're not going to stick around this place and not take bets."[10]

Starting in 2004, GE has begun to shift its "operating model" from almost a singular focus on efficiency and effectiveness to a balanced focus on efficiency and innovation. With significant commitment from the CEO, over the last seven years GE has changed its external image from a firm focused on efficiency to a firm focused on innovation. Look no further than the way GE positions itself and its focus on ecomagination. Ecomagination is GE's initiative to meet customer demands for more energy efficient products while driving growth for GE. GE has demonstrated the ability to grow organically using innovation to drive new products and services, and it has topped its goals for organic growth in the last three years, while retaining its focus on efficiency and quarterly results. This transition took seven years, significant dedication from the executive team, and a major shift of perspectives, attitudes, and resources within the organization.

Was the transition painless? No. Many executives and middle managers found the change in BAU models difficult. One consultant quoted in the *BusinessWeek* story said "It seems painful to them, like a waste of time." Paul Bossidy, who was then the CEO of GE Commercial Equipment Financing, said "This is a big fundamental structural change, and that can be tough." The transition to a more innovative culture encountered another form of resistance—the recruitment of experienced outsiders. GE had traditionally hired from within, but Immelt recognized that new skills and new perspectives were important. The *BusinessWeek* article called the increase in external recruitment a "gut punch to the culture" for GE.

So the transition hasn't been smooth or easy, and it hasn't been particularly fast, but it has demonstrated significant value. The shift from an overwhelming focus on efficiency and effectiveness to a balanced focus on both efficiency and innovation has been beneficial for GE and continues to this day, almost eight years after the concept of the Imagination Breakthrough was announced. GE, however, isn't finished with the transition.

Results

After lauding GE for its focus on rebalancing its operating model, it does make sense to evaluate the results. The results to date for GE have been mixed. Immelt has clearly communicated and pursued a focus on innovation. In GE's 2010 annual report, Immelt focuses on GE's ability to "innovate at a large scale" and promises to launch more than 100 health care innovations in the coming fiscal year.[11] GE has recovered from the depths of the financial meltdown, but its stock price still lags its closest competitors. In an article for *Fortune* magazine dated February 10, 2011, Geoff Colvin suggests that GE has underperformed the market and its competitors over the last decade. Some of that is attributable to the fact that GE was exposed to the financial crisis in ways that its competitors, such as Siemens, weren't. Colvin points out that GE has done a poor job picking winners and losers and perhaps it isn't doing as well generating strong executive management.[12] In a firm as large and diverse as GE, with a historic focus on efficiency, shifting the BAU to a balance between innovation and efficiency will take time. Clearly, GE is in the midst of a transition, and time will tell if its work to rebalance the operating model will pay dividends.

Balance for Survival

Google, Procter & Gamble, Gore, Apple, and 3M, the firms I call relentless innovators, have few attributes in common. They compete in different markets with different product or service offerings. Procter & Gamble and 3M are relatively old, having been around for at least 100 years. Google and Apple are relatively new firms. Yet these firms demonstrate what I believe is the only real sustainable competitive advantage—they innovate constantly, relentlessly.

These firms have all the characteristics I've identified in this chapter that are important for innovation success:

- Committed senior executives
- Openness to new tools and new thinking
- An emphasis on innovation as a revenue driver
- Nimbleness
- A rapid experimentation culture
- The ability to develop ideas over time
- A senior executive who created or creates "Cortes" moments—for example, A. G. Lafley stating that 50 percent of innovations would come from external

sources, or Steve Jobs introducing new innovations and constantly resetting the bar for Apple.

These attributes exist as part of the culture and expectation of the business—they aren't dependent on one visionary leader or a market crisis.

Given the importance of relentless, continuous innovation for survival, your firm must adopt these characteristics as part of its BAU or prepare to defend against those that do. I'm not suggesting that the shift from efficiency to innovation will be easy. GE demonstrates that it is hard work and it takes time. I am suggesting that the shift is vital for success in the future, as innovation becomes increasingly important for growth, differentiation, and survival.

Chapter 10

Results of the Shift

What are the results of shifting your operating model toward a balance between efficiency and innovation? What happens when your middle managers have the skills necessary to achieve this balance in day-to-day operations?

Everything changes.

First, perspectives and attitudes change. When new challenges arrive or problems crop up, managers reach for a range of tools and techniques, rather than simply deciding to force more efficiency out of internal processes. In many organizations innovation tools and techniques are used only in emergencies, when all other approaches have failed. After the transition to an innovation BAU model, innovation tools and techniques are deployed consistently, in all situations. Increasingly the organization becomes more proactive, attuned to scanning the horizon, watching and assessing trends, and anticipating emerging opportunities or threats. This shift means less firefighting is necessary and more time and resources can be brought to bear on new products, services, and business models.

As part of that new perspective, managers and executives expect and anticipate innovation, viewing it as a business discipline that can be expected to generate results as consistently as other defined operating models. This new-found confidence in innovation as a discipline reduces concerns about disruption of BAU. Clearly, the renewed balance doesn't relieve the firm of the requirement to achieve predictable quarterly results, but now managers are much more aware of opportunities for innovation and the resulting benefits. Whereas MM historically focused the vast majority of their time and resources in keeping the core business humming, now middle managers identify instances, challenges, and market opportunities in which innovation can play a critical role in advancing the business. The successful middle managers, and some executives, increasingly become multidimensional.

These shifts are only possible if the culture, attitudes, perspectives, processes, and methods that comprise "business as usual," both the formal, stated concepts and the informal, unstated methods, embrace innovation as a consistent business process and discipline. This shift means that every factor in the BAU arsenal is now attuned to accept innovation as a consistent capability. Every team, geography, and product group is focused on a balance between efficiency and innovation.

A New Management Disruption

Over the last 20 years businesses have faced a number of significant disruptions, usually caused by external factors or shifts that forced managers and operating models to change. For an example, consider the Y2K problem. As the modern world became more dependent on computers, it was unthinkable that a date problem would cause major infrastructure systems to collapse, yet the year 2000 date challenge left many IT organizations scrambling to replace or upgrade computer systems. Y2K accelerated the adoption of new computing systems and technologies, which probably would not have been adopted with so much alacrity otherwise. New systems, installed to avoid the potential Y2K disaster, changed the way many companies work, driving unexpected benefits, such as tighter integration or higher efficiency or lower costs. Other disruptions, like the growing adoption of the Internet as a marketing and sales channel, or the increasing adoption of free trade, demonstrate that many significant shifts are led by forces from external to the organization, requiring the firm to respond.

Balancing innovation and efficiency, however, is different. There's no external market shift, no impending Y2K disaster, no significant technological introduction. Shifting the business model to create a balance between innovation and efficiency must start from the inside, based on a firm's needs for growth and differentiation. The changes must take place first in the operating model, shifting the way people manage, which will shift internal perspectives. In many ways this change is more difficult than dealing with external disruptions, since it isn't based on outside threats or impending technological disasters, but on the slow erosion of market share and relevance.

Other shifts, like the wholesale adoption of ERP systems in the late 1990s in response to the Y2K problem, used a crisis to radically alter internal operating models. Previous disruptions to operating models were caused by dramatic shifts in the market or impending deadlines. The requirement to balance efficiency and innovation in an internal operating model has much more subtle signals, indicated by decreasing profitability, reduced market share, and stagnation. The signals are not as persistent or overt, however. Firms that fail to change will wither slowly, the death of 1,000 cuts. The shift to rebalance the operating model must originate and be sustained internally, against internal resistance and inertia. Starting this shift won't be easy, but once you have your "operating model" rebalanced and your middle managers on board, your firm can accomplish amazing things.

Take on the BHAGs

Jim Collins and Jerry Porras coined the phrase BHAG, a Big Hairy Audacious Goal, in an article titled "Building Your Company's Vision."[1] A BHAG is simply a compelling, powerful long-term goal that your firm sets for itself. Once your operating model is balanced between efficiency and innovation, your firm can define and achieve its own BHAGs (consider P&G's 50 percent goal). Any firm can declare a BHAG, but if the operating model is focused on efficiency at the expense of innovation, those goals can't be achieved. Consider these three questions in relation to your company:

1. What are the BHAGs in your business?
2. What is within reach if only your operating model was in balance and your teams had the skills and resources to innovate more successfully?
3. If you don't reach for those BHAGs, what competitor or new entrant will?

Too often innovation solves only "small bore" problems because the operating model won't tolerate the risk and uncertainty required to take on larger innovation goals. Only firms that effectively balance efficiency and innovation can create the revenues to fund innovation and tolerate the risks necessary for the BHAGs. As the innovation "business-as-usual" operating model comes online, vision and expectations increase. The scope and nature of corporate and industry challenges increase. The tasks executives set for their firms become far larger. Big, hairy goals become game-changers that force your competitors to react to your products and services.

Make Innovation a Business Discipline

If your operating model is balanced between efficiency and innovation, innovation becomes an expected, "everyday" capability or discipline that is consistently exercised, rather than an occasional, uncertain, and difficult experiment. This shift means every opportunity, issue, and threat is examined from two perspectives:

- How can your firm be more efficient and/or effective to meet this issue or opportunity?
- How can your company use innovation to meet this opportunity or challenge?

Managers and executives will have more tools and information at their disposal and they will be prepared to respond appropriately from both perspectives; the benefits are obvious. Your team gains the capability to examine any problem from multiple perspectives, rather than merely considering how to eliminate costs. Your team may discover entirely new revenue streams or business models from challenges that were once considered dire threats. Further, as innovation becomes a discipline, new opportunities will be uncovered much more frequently, with more viability, and they will be converted into new products and profits with greater consistency.

Use All Innovation Tools Effectively

While your business model is heavily weighted toward efficiency and effectiveness, your team may acquire a number of innovation tools, techniques, and methods, but it will struggle to implement them completely. Your teams will never be satisfied with the results obtained when using these tools if the model is heavily focused on efficiency. This problem is not with the tools or techniques, but with the capability of your middle managers to successfully deploy innovation in the face of a resistant operating model. Only when your operating model is balanced and encourages efficiency and innovation together will you be able to deploy innovation tools and methods effectively and gain the full benefit of their use.

Whether your team favors disruptive innovation, "needs-based" innovation, TRIZ, or "open" innovation, the desired results will be attainable once the operating model comes into balance. Innovation tools, whether process-based like Systematic Inventive Thinking, or those focused on specific outcomes, like disruptive innovation, must align with and mesh with the predominant operating model in order to be effectively deployed. Once the operating model is tuned to expect innovation, and it is adapted to integrate and use those tools, innovation is simply accelerated.

The same claim is true for the use of idea management tools. What many firms discover when using idea management tools is an initial outpouring of ideas that have been pent up for years, followed by less interaction and engagement. If the BAU is balanced between efficiency and innovation, the idea management software becomes more than just an ideas database—it becomes an idea management application supporting, enabling, and scaling the innovation capability.

Your team can acquire and use any innovation tool, technique, or methodology, but the results won't be truly exceptional until the internal operating model is balanced between efficiency and innovation. Once the model is well balanced, and middle managers are supportive of innovation, the outcomes from excellent innovation tools and idea management systems will be magnified.

Stop Fighting Fires and Cause a Few

Once the operating model is balanced, your firm can stop reacting to sudden changes created by regulations, legislation, demographics, and technology, and begin to anticipate these shifts. Trends and scenarios are used to anticipate change and spot emerging threats or opportunities before others do. Using that knowledge, your firm can stop "fighting fires" based on new products introduced by competitors or new market disruptions. Rather, your team can become proactive, introducing new products and services that disrupt the existing markets, which force your competitors to react. Stop reacting to market conditions and start causing your competitors to become firefighters.

Now that your firm has taken the initiative, your competitors have no choice but to react to your leadership. With enough foresight, your firm will progress further, taking a new position, producing new products, or shifting the business model just as rival companies think they've "caught up." Apple, for example, has become a master at forcing firms that manufacture cell phones, MP3 players, and tablets to play "catch up," and it is always one step ahead. Curiously, while Apple is recognized as the innovation leader in these categories, it is also the efficiency leader, as witnessed by the cost models and pricing for the iPad versus its competition.[2] Apple demonstrates that the operating model can be effectively balanced between efficiency and innovation—a firm can focus and "win" on both counts, as opposed to over-relying on one capability and dismissing or ignoring the other.

Innovation Becomes a "Natural" Process

In many firms, innovation seems disjointed, awkward, and sporadic, because everyone recognizes the ideas don't mesh with the expectations and process. Since the operating model is well defined, broadly accepted, and consistent over time, innovative ideas that don't align to current business expectations and processes are rejected. In a business that has an appropriate balance between efficiency and innovation, innovation seems like a natural process. Ideas are routinely generated and reviewed by people who are accustomed to evaluating ideas. Ideas are encouraged and accepted and they are managed in a fluid process. This fluidity means that innovation doesn't require your team to start suddenly, reacting to competitive pressures with people who don't know their roles and work with ill-defined processes. Innovation works as effortlessly and as seamlessly as other well-defined transactional processes, guided and supported by engaged middle managers. In a firm with a balanced operating model, ideas are expected, encouraged, and managed effectively in the operating model.

> Within an innovation "business-as-usual" operating model, innovation is an effective, efficient process, sustained by the very middle managers who previously would have rejected the concept!

Leverage Your Networks More Effectively

While clearly not the primary focus, rebalancing your internal operating model will help leverage your existing partnerships and perhaps create new ones. Once your operating model is balanced, your team will identify innovation opportunities that originate inside the organization, and increasing opportunities that arise from customers and partners. Almost every relentless innovator has embraced open innovation but could not have been successful leveraging their partnerships without robust internal innovation capabilities. As your operating model focuses on innovation as business as usual, middle managers will identify the best source of new ideas, including internal and external sources. Open innovation becomes another tool at the disposal of executives and middle managers.

Middle Managers Truly Are Your Best Assets

The phrase "our employees are our best assets" is a cliché, but middle managers who learn to stretch their skills and embrace a new operating model balanced between efficiency and innovation will increase their skills, add tremendous value to the bottom line, and become the most valuable resource in the organization. Middle management is capable of cutting costs and creating greater efficiencies on one hand, or creating new products and services that drive new revenue, profits, and differentiation on the other hand. With an innovation BAU, these individuals will find more excitement and growth at work, since they will be able to exercise more than one set of skills. Their enthusiasm for their employer and job will increase as well.

Human resource executives are constantly seeking the "holy grail" of employee engagement. However, a focus on efficiency and cost-cutting doesn't encourage such an environment. Few people plan to make their life's work a never-ending pursuit of greater efficiency. Instead, managers pursue efficiency in larger organizations because they don't believe they have any other option. Most middle managers do not have a great love of efficiency and would far rather identify new opportunities and develop new products and services. They pursue efficiency because they have to, and the work doesn't inspire their engagement or their passion. People who work in jobs that allow them to pursue their passions and interests, however, are much more highly engaged.

As your operating model shifts from a focus on efficiency to a focus on innovation, and a balance between the two, your human resources team won't need special programs to encourage more engagement. The nature and type of the work under a more consistent innovation focus will allow people to pursue their interests and passions, which automatically will lead to higher engagement. Firms that we hold up as leading innovators already understand this. It's no surprise to find that Apple, Google, 3M[3] and Gore[4] have significantly lower unplanned employee turnover than their competitors and higher levels of employee engagement.

Conclusion

In the end, innovation isn't about intricate new techniques or philosophies. It isn't dependent on "open innovation" or overly creative people. Relentless innovators understand that innovation is a core competency that requires defining an innovation business-as-usual operating model, then sustaining and enabling that BAU competency with the most effective and capable people in your organization.

Ultimately, consistent innovation success is about a shift from *efficiency as business as usual* to *innovation as business as usual*. Innovation success relies on making innovation something that people expect and anticipate, rather than an occasional, unusual exercise. Relentless innovators understand this, and this book describes how your company can change your operating model and refocus your team to become relentless innovators.

More than 20 years ago, psychologist Mihaly Csikszentmihalyi wrote a book entitled *Flow: The Psychology of Optimal Experience*. In the book he notes that many creative people are able to achieve a state in which their creativity seems to flow, almost naturally, and they lose track of time, obstacles, and other barriers. Individuals who achieve "flow" can be exceptionally productive while in that state, sparking far more creativity than otherwise. In *Relentless Innovation*, I've attempted to demonstrate how a firm can achieve "innovation flow" by rebalancing the operating model between efficiency and innovation, and by refocusing middle managers and encouraging them to embrace innovation.

Most firms experience outcomes far different from the "innovation flow" I've described. Their innovation initiatives are difficult, distracting, sporadic, and rarely successful. Increasingly, innovation is viewed with skepticism and has become synonymous with such frustration and failure. I've postulated that the reason innovation initiatives in many firms are so far removed from the conditions Csikszentmihalyi terms "flow" is that the two factors that drive so much success in your business—business as usual and middle management—hinder innovation.

Rebalancing the operating model to incorporate both efficiency and innovation will encourage more innovation activity, but only if the middle managers have the skills and incentives to attempt more innovation. If both the operating model and middle managers are encouraged and permitted to innovate, your organization has a much better chance to achieve innovation flow, remaining viable and competitive, while optimizing financial success. Because innovation is no longer a nice to have, it is a must have.

Notes

Chapter 1

1. http://www.fastcompany.com/blog/david-brier/defying-gravity-and-rising-above-noise/life-branding-needs-vision-too
2. http://www.nokia.com/about-nokia/company/story-of-nokia
3. http://www.foxbusiness.com/markets/2011/02/09/nokia-market-share-slides-gartner
4. http://www.helsinkitimes.fi/htimes/domestic-news/business/12634-qcomplacentq-nokia-killed-touchscreen-handset-in-2004-nyt-.html
5. http://www.engadget.com/2011/02/08/nokia-ceo-stephen-elop-rallies-troops-in-brutally-honest-burnin

Chapter 2

1. http://www.fool.com/investing/general/2011/02/25/apples-management-is-creating-value.aspx
2. Dyer, Gregersen, and Christensen, *The Innovator's DNA: Mastering the Five Skills of Disruptive Innovation,* Cambridge, MA: Harvard Business Press, 2010, pp. 160–161.
3. http://www.forbes.com/special-features/innovative-companies-list.html

Chapter 3

1. http://www.nist.gov/baldrige/about/history.cfm
2. http://www.time.com/time/nation/article/0,8599,1868997,00.html
3. http://www.nsf.gov/statistics/infbrief/nsf11300/nsf11300.pdf
4. http://www.nsf.gov/statistics/infbrief/nsf11300/nsf11300.pdf
5. http://www.advfn.com/news_3M-CEO-Innovation-Fueling-Company-s-Strong-Performance_42768477.html
6. http://www.sas.com/company/annual-report-current.pdf
7. http://www.nsf.gov/statistics/infbrief/nsf11300/nsf11300.pdf
8. Geoff Tennant, *Six Sigma*, page 6.
9. John F. Krafcik, (1988), "Triumph of the lean production system," *Sloan Management Review*, 30 (1): 41–52.

Chapter 4

1. http://web1.ncaa.org/stats/StatsSrv/ranksummary
2. http://www.nba.com/draft/2011
3. http://hoopshype.com/salaries.htm and author's calculations
4. http://nces.ed.gov/programs/digest/d10/tables/dt10_206.asp?referrer=list
5. http://nces.ed.gov/programs/digest/d10/tables/dt10_341.asp
6. http://stocks.investopedia.com/stock-analysis/2009/invest-in-the-original-sp-500-gm-lo-pep-cl-cr0218.aspx
7. http://www.businessweek.com/innovate/content/feb2011/id20110225_141049_page_2.html
8. http://www.zdnet.com/blog/apple/aapl-tops-300-still-waiting-on-michael-dells-apology/8439
9. Henry Chesbrough, *Open Innovation: The New Imperative for Creating and Profiting from Technology*, Cambridge, MA: Harvard Business Press, 2005.

10. I provide an Open Innovation Typology in Chapter 3 of *A Guide to Open Innovation and Crowdsourcing: Advice from Leading Experts*, edited by Paul Sloane. London: Kogan Page, 2011.
11. To learn more about defining an innovation process, see my book entitled *Make Us More Innovative: Critical Factors for Innovation Success*, Jeffrey Phillips, 2008, or *The Other Side of Innovation: Solving the Execution Challenge*, by Vijay Govindarajan and Chris Trimble, Cambridge, MA: Harvard Business Press, 2010.

Chapter 5

1. http://hbr.org/1999/09/creating-breakthroughs-at-3m/ar/1

Chapter 6

1. http://www.lockheedmartin.com/aeronautics/skunkworks

Chapter 7

1. http://www.businessweek.com/magazine/content/06_05/b3969401.html
2. http://www.ft.com/intl/cms/s/0/c73ea5cc-11e9-11e0-92d0-00144feabdc0.html?ftcamp=rss#axzz1RQFji3LT
3. http://www.ideaconnection.com/interviews/00055-Innovation-and-Tobacco-at-R-J-Reynolds.html
4. Charles O'Reilly III, J. Bruce Harreld, Michael Tushman "Organizational Ambidexterity: IBM and Emerging Business Opportunities," *Stanford Graduate School of Business Research Paper No. 2025; Rock Center for Corporate Governance Working Paper No. 53*, 2009.
5. O'Reilly et al, page 25.

6. O'Reilly et al, page 26.
7. O'Reilly et al, page 26.
8. O'Reilly et al, page 26.
9. http://www.newyorker.com/reporting/2011/05/16/110516 fa_fact_gladwell
10. Peter Schwartz, *The Art of the Long View: Planning for the Future in an Uncertain World*; New York: Currency Doubleday, April 15, 1996.

Chapter 8

1. Rosabeth Moss Kanter, *The Change Masters*. New York: The Free Press, 1985.
2. Dyer, Gregersen, and Christensen, *The Innovator's DNA: Mastering the Five Skills of Disruptive Innovators,* Cambridge, MA: Harvard Business Press, 2010.
3. Johnson, *Natural History of Innovation,* page 166.

Chapter 9

1. http://abcnews.go.com/Technology/story?id=4839327& page=1
2. http://www.businessweek.com/magazine/content/05_40/ b3953093.htm
3. http://query.nytimes.com/gst/abstract.html?res=F40C1EFC3 95F0C728CDDAB0894DD404482
4. http://frontendofinnovation.blogspot.com/2011/05/fei2011 -morning-learnings.html
5. Peter Sims, *Little Bets: How Breakthrough Ideas Emerge from Small Discoveries*, New York: Free Press, 2011.
6. http://hbr.org/product/ge-s-growth-strategy-the-immelt -initiative/an/306087-PDF-ENG

7. http://www.mbanerds.com/index.php?title=GE%27s
_Growth_Strategy:_The_Immelt_Initiative
8. Business 2.0, July 2004, GE sees the light.
9. "The Immelt Revolution," *BusinessWeek,* March 28, 2005.
10. Ibid.
11. http://www.ge.com/ar2010/letter.html#!letter=page-1
12. http://management.fortune.cnn.com/2011/02/10/grading
-jeff-immelt

Chapter 10

1. James C. Collins and Jerry I. Porras, "Building Your Company's Vision," *Harvard Business Review*, 65, No. 1, October 1996.
2. http://www.marketwatch.com/story/apples-ipad-price
-advantage-the-breakdown-2011-03-02
3. Charles Hill and Gareth Jones, *Essentials of Strategic Management*, South-Western, page 76.
4. http://money.cnn.com/magazines/fortune/bestcompanies
/2011/turnover

Recommended Readings

Any valuable area of practice has a defined body of knowledge. Innovation is no different. Many leading innovation thinkers have documented their concepts in excellent books that comprise a vast body of knowledge about innovation, and I'd be remiss if I didn't identify some of the best of those books, which have informed my thinking and should inform yours.

Much of the innovation work we do today is grounded in the work of the "godfather" of creativity and innovation, Alex Osborn. His book *Applied Imagination*, published in 1963, coined the phrase "brainstorming" and should be a reference book on any innovator's desk.

Other books I refer to regularly and recommend to my clients, regardless of their innovation need, are:

Henry Chesbrough, *Open Innovation: The New Imperative for Creating and Profiting from Technology*, Cambridge, MA: Harvard Business Press, 2005.

- The primer on "open" innovation

Clayton Christensen, *The Innovator's Dilemma: When New Technologies Cause Great Firms to Fail*, Cambridge, MA: Harvard Business Press, 1997.

- Reignited the corporate focus on innovation

Jeff Dyer, Hal Gregersen, Clayton M. Christensen, *The Innovator's DNA: Mastering the Five Skills of Disruptive Innovators*, Cambridge, MA: Harvard Business Press, 2010.

- Defines the five key attributes or skills that all good innovators possess

Vijay Govindarajan and Chris Trimble, *The Other Side of Innovation: Solving the Execution Challenge*, Cambridge, MA: Harvard Business Press, 2010.

- Identifies the key challenge to innovation: defining an execution strategy to convert concepts into new products or services

Gary Hamel, *The Future of Management*, Boston: Harvard Business School Publishing, 2007.

- Introduces the idea of "management" or organizational hierarchy innovation

Andrew Hargadon, *How Breakthroughs Happen*, Boston: Harvard Business Press, 2003.

- Defines the innovation approach for disruptive innovation

Tim Hurson, *Think Better: An Innovator's Guide to Productive Thinking*, New York: McGraw-Hill, 2008.

- Provides the tools and insights to help anyone become a better innovation facilitator

Tom Kelley, *The Ten Faces of Innovation*, New York: Doubleday, 2005.

- Examines the key roles that support innovation

Alex Osterwalder and Yves Pigneur, *Business Model Generation*, Hoboken, NJ: Wiley, 2010.

- Defines a methodology for innovating a business model

Keith Sawyer, *Group Genius*, New York: Basic Books, 2008.

- Examines methods to achieve more innovation from a group

Peter Schwartz, *The Art of the Long View: Planning for the Future in an Uncertain World*, New York: Currency Doubleday, 1996.

- The primer on trend spotting and scenario planning

Peter Sims, *Little Bets: How Breakthrough Ideas Emerge from Small Discoveries*, New York: Free Press, 2011.

- Describes how to create a culture of experimentation in your organization

Eric von Hippel, *Democratizing Innovation*, Boston: MIT Press, 2005.

- Introduces the idea of customer-centered or user-centered innovation

Roger von Oech, *A Whack on the Side of the Head*, US Games Systems, Inc., 1990.

- One of the best books to spark creativity ever written

Index

About the Author

Jeffrey Phillips leads the innovation consulting team at OVO Innovation. OVO works with Fortune 500 organizations to create innovation capacities by defining and building innovation teams, designing and developing internal innovation processes, and developing open innovation skills and capabilities. OVO has developed successful innovation programs for firms in a range of industries, including financial services, high technology, insurance, software, consumer goods, pharmaceuticals, and government agencies.

Jeffrey writes the popular Innovate on Purpose blog, providing insights and perspectives into innovation methods, tools, and challenges. He is the author of *Make Us More Innovative*, a book that defines OVO's Innovate on Purpose methodology Jeffrey has also contributed to several other books on innovation, including *A Guide to Open Innovation and Crowdsourcing*, edited by Paul Sloane. He is regularly published on innovation sites and management journals such as InnovationTools .com, InnovationManagement.se, InnovationExcellence.com, *Texas Enterprise*, and many more.

Jeffrey is a recognized speaker on innovation topics, and has led innovation talks, workshops, and programs in North America, Europe, the Middle East, and Southeast Asia. He has led talks for corporate boards, industry consortia, innovation conferences, and in a number of major universities.